Carl Friedländer

**The Use of the Microscope in Clinical and Pathological Examinations**

Carl Friedländer

**The Use of the Microscope in Clinical and Pathological Examinations**

ISBN/EAN: 9783744690218

Printed in Europe, USA, Canada, Australia, Japan

Cover: Foto ©berggeist007 / pixelio.de

More available books at **www.hansebooks.com**

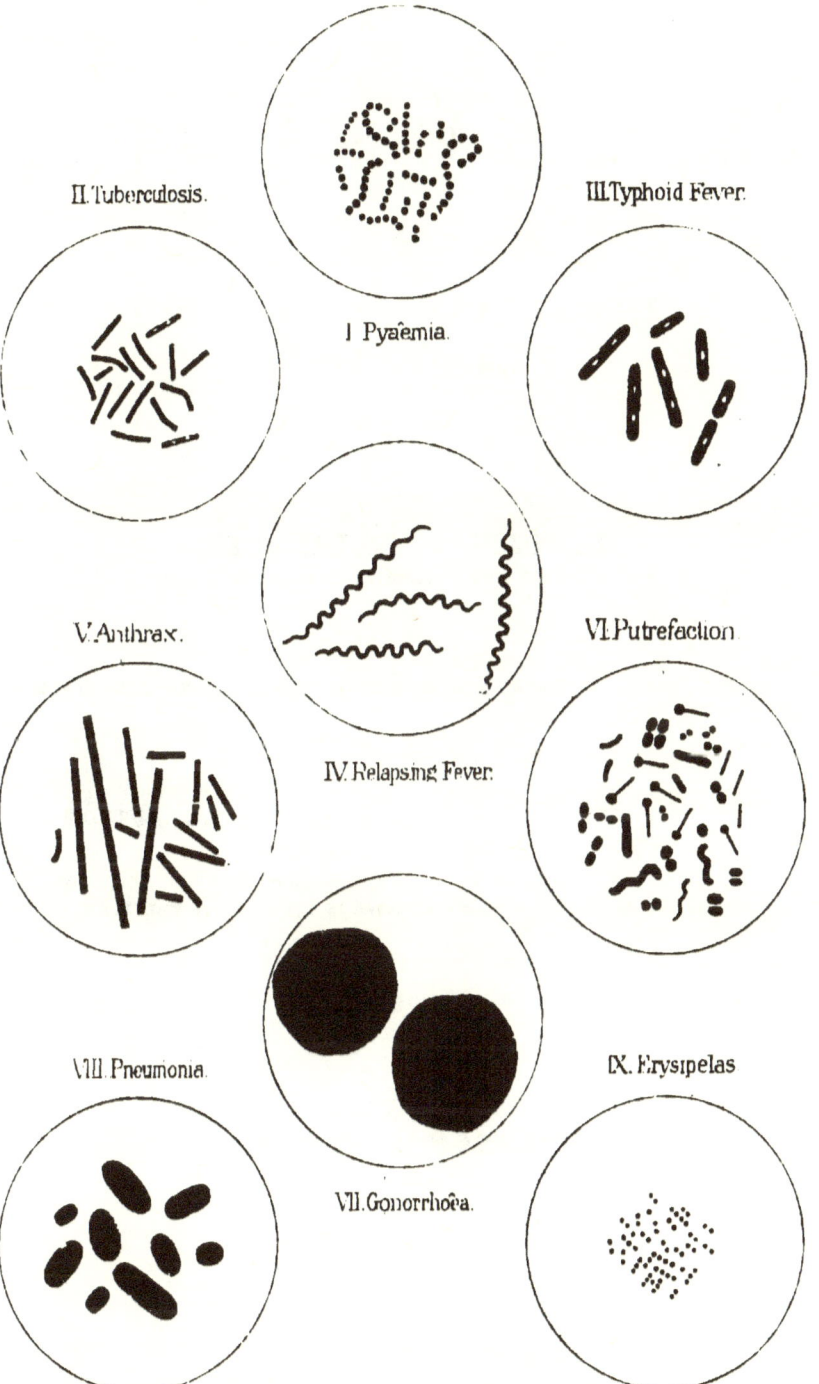

# THE USE

OF THE

# MICROSCOPE

IN

## CLINICAL AND PATHOLOGICAL EXAMINATIONS.

BY

Dr. CARL FRIEDLAENDER,
PRIVAT-DOCENT IN PATHOLOGICAL ANATOMY AT BERLIN.

SECOND EDITION, ENLARGED AND IMPROVED, WITH A CHROMO-LITHOGRAPH.

*TRANSLATED, WITH THE PERMISSION OF THE AUTHOR,*
By HENRY C. COE,
M. D., M. R. C. S., L. R. C. P. (LONDON),
PATHOLOGIST TO THE WOMAN'S HOSPITAL IN THE STATE OF NEW YORK.

NEW YORK:
D. APPLETON AND COMPANY,
1, 3, AND 5 BOND STREET.
1885.

COPYRIGHT, 1885,
BY D. APPLETON AND COMPANY.

# PREFACE TO THE FIRST GERMAN EDITION.

The author has been frequently requested to give a brief description of those methods which are adopted in microscopical examinations, conducted for diagnostic and pathological purposes. These same methods, which twenty years ago were extremely simple, have gradually become more complicated and have, in many respects, been very essentially improved and refined. A large number of the surprising advances which have been made of late years in the field of vegetable parasites, are the direct results of improvements in *technique*.

If, therefore, every one who aspires to follow the progress of pathology must become acquainted with recent methods, this is still more incumbent upon him who desires to undertake microscopical examinations for clinical and pathological ends. A concise statement of the processes employed in pathological histology has hitherto been wanting, and its absence has been deeply felt by many; this book aims at supplying the deficiency. The fact that examinations for schizomycetes have been treated quite exhaustively, and have received especial preference, ought to meet with universal approval. It has frequently been necessary to enter rather deeply into the diagnostic and prognostic significance of discoveries; particular stress was laid upon this in discussing the subject of tubercle-bacilli in sputa, and in deciding between erosion and carcinoma of the uterus.

May this little volume serve as a guide to the beginner in this study, which is as fascinating as it is difficult. Perhaps the expert, too, may find here and there a useful hint.

CARL FRIEDLAENDER.

CITY HOSPITAL, BERLIN, *August*, 1882.

# PREFACE TO THE SECOND GERMAN EDITION.

A SECOND edition of this little book has become necessary after the lapse of a comparatively short period. I have aimed to select from the many new contributions on technology, and to add to the text, those which present essential improvements as regards my purposes. I have also, in accordance with the wishes of many, presented in the appended colored plate a comparison of the most important and characteristic pathogenic schizomycetes. I am indebted for the drawings to my honored friend and fellow-worker Dr. Gram, of Copenhagen. They have all been executed on the same scale—one to one thousand.

THE AUTHOR.

BERLIN, *April*, 1884.

# CONTENTS.

## CHAPTER I.
### THE MICROSCOPE.
|   |   |
|---|---|
| 1. The stand. Abbé's apparatus | 1 |
| 2. Objectives. Water- and oil-immersion lenses | 4 |
| 3. Eye-pieces. Accessory apparatus. Combinations | 7 |

## CHAPTER II.
### ACCESSORIES.
1. Illuminating lamp . . . . . . . . . 8
2. Glass apparatus . . . . . . . . . 9
3. Metallic instruments . . . . . . . . . 10
4. The microtome. Thick and thin sections; further treatment of sections . . . . . . . . . . . 11

## CHAPTER III.
### REAGENTS. MICRO-CHEMISTRY.
Artificial products . . . . . . . . . 19
Micro-chemical examinations . . . . . . . 21
1. Distilled water . . . . . . . . . 22
2. Salt-solution of 0·8 per cent. Indifferent fluid . . . . 23
3. Absolute alcohol. Hardening . . . . . . 24
4. Ether and chloroform. The removal of fat . . . . 27
5. Acids . . . . . . . . . . . 28
    *a.* Sulphuric, hydrochloric, and nitric acids. Decalcification . 28
    *b.* Acetic acid . . . . . . . . . 30
    *c.* Picric acid . . . . . . . . . 33
    *d.* Chromic acid; chromates. Müller's fluid . . . . 34
6. Alkalies. Liquor potassæ and liquor sodæ. Ammonia . . 36

## CONTENTS.

|   |   |
|---|---|
| 7. Glycerin | 38 |
| 8. Acetate of potassium | 40 |
| 9. Oil of cloves, Canada balsam | 40 |
| Reagents used in staining. The principles of staining | 42 |
| 10. Iodine | 46 |
| Glycogen | 47 |
| Corpora amylacea | 48 |
| Amyloid material | 48 |
| 11. Carmine | 51 |
|     *a.* Ammonia-carmine | 51 |
|     *b.* Picro-carmine | 54 |
|     *c.* Borax-carmine | 56 |
|     *d.* Alum-carmine | 56 |
|     *e.* Cochineal-alum solution | 57 |
|     *f.* Lithium-carmine | 57 |
| 12. Hæmatoxylin | 58 |
| Weigert's method of staining nerve-fibers | 58 |
| 13. Eosin | 60 |
| 14. Aniline-black (nigrosin). Aniline-blue | 62 |
| Staining of nuclei | 62 |
| Food-cells | 66 |
| Staining of amyloid substance | 66 |
| 15. Identification and staining of schizomycetes | 67 |
|     *a.* Identification of schizomycetes when unstained | 68 |
|     *b.* Staining of micrococci | 71 |
|     *c.* Gram's method | 75 |
|     *d.* Staining of tubercle-bacilli | 78 |
| 16. The noble metals. | |
|     *a.* Silver | 86 |
|     *b.* Gold | 88 |
|     *c.* Osmic acid | 89 |
| 17. Ammonium sulphide | 90 |

### CHAPTER IV.
#### OTHER METHODS OF PREPARATION.

|   |   |
|---|---|
| 1. Boiling | 93 |
| 2. Drying | 94 |
| 3. Artificial digestion | 94 |
| 4. Imbedding. Celloidine | 97 |
| 5. The process of injection | 100 |
|     *a.* Injecting materials | 100 |
|     *b.* Injecting apparatus | 103 |
| 6. Preservation of specimens | 106 |

# CONTENTS.

## CHAPTER V.
### THE OBSERVATION OF LIVING TISSUES.

The circulation. Inflammation . . . . . . . 109
1. The frog's web . . . . . . . . . . 110
2. The tongue . . . . . . . . . . 110
3. The mesentery . . . . . . . . . . 111
4. The cornea . . . . . . . . . . 112

## CHAPTER VI.
### THE EXAMINATION OF FLUIDS.

Vital properties of the suspended elements. Amœboid movements 114
The form of the elements . . . . . . . . 116
Examination of tissue-fluid . . . . . . . . 117
Examination of micro-organisms . . . . . . . 118
Koch's method of staining dried preparations . . . . . 121
1. Blood . . . . . . . . . . . 126
   *a*. Diminution of the number of red blood-corpuscles in anæmia 127
   *b*. Change in the size and shape of the red blood-corpuscles. Nucleated red blood-corpuscles . . . . . . 128
   *c*. Increase in the number of white blood-corpuscles. Leucocytosis and leucæmia. Changes in the granular protoplasm . 129
   *d*. Other cell-elements in the blood. Worms and schizomycetes 131
   *e*. The examination of blood-stains. Hæmin-crystals. Hæmatoidin . . . . . . . . . . . 134
2. Sputa . . . . . . . . . . . 136
   *a*. Oral fluids . . . . . . . . . . 137
   *b*. Products of the respiratory mucous membrane . . . 141
   *c*. Elastic fibers. Fibrinous exudations. Asthma-crystals . 145
   *d*. Schizomycetes. Tubercle-bacilli. Pneumococci . . . 146
3. Pus.
   *a*. Pus-corpuscles and fatty degenerated cells . . . . 155
   *b*. Foreign substances . . . . . . . . 156
   *c*. Schizomycetes and actinomycetes . . . . . . 158
4. Urine.
   *a*. Precipitates and crystals . . . . . . . 159
   *b*. Casts . . . . . . . . . . . 161
   *c*. Pus- and mucus-corpuscles. Epithelial cells . . . . 163
   *d*. Tumor-elements . . . . . . . . . 164
   *e*. Entozoa . . . . . . . . . . 164
   *f*. Vegetable parasites . . . . . . . . 164

5. Secretions of the genital tract.
   *a.* The vaginal secretion . . . . . . . 166
   *b.* Fluids from the uterus . . . . . . . 167
       Carcinoma or erosion . . . . . . . 169
   *c.* Gonorrhœal secretion . . . . . . . 173
   *d.* Semen and the prostatic secretion . . . . 174
6. Contents of the stomach and intestine.
   *a.* Remains of food . . . . . . . . 175
   *b.* Epithelial cells, mucus, etc. . . . . . . 176
   *c.* Entozoa . . . . . . . . . . 177
   *d.* Vegetable parasites . . . . . . . 177
7. Exudations. Contents of cysts . . . . . . 179

## CHAPTER VII.

The examination of solid elements of the body, extirpated tumors, etc. . . . . . . . . . . . 183

THE

# USE OF THE MICROSCOPE.

## I.

### THE MICROSCOPE.

IN regard to the choice of a microscope, the principle is to be established that the lenses and the stage must be free from defects. Do not be led to purchase an instrument of inferior quality, because the price is a little lower; most of the objects which we have to examine with the microscope present so many difficulties, their outlines are so delicate, etc., that it is only under the most favorable circumstances that we succeed in bringing them into view perfectly, and without great loss of time. Therefore, it is best to select at the outset an instrument of a well-known and reliable make. If one of the lenses, or other parts, which have been furnished, does not satisfy *all* of the proper requirements, send them back at once, for nothing is more unpleasant than a year's struggle with an imperfect microscope.

On the contrary, it is by no means necessary to purchase the strongest objectives at the start, since they materially increase the price of instruments. It is better for the beginner to use only low and moderate powers at first—three hundred at the highest;

the management of stronger lenses is beset with so many difficulties, and demands such extreme accuracy, that it is decidedly advisable to prepare one's self for this by previously working for some time with lower powers.

1. THE STAND (Abbé's Apparatus).—The stand must be so arranged that it can be used even with the strongest objectives; above all, the action of the micrometer-screw must be sufficiently delicate. The stage must be large, firm, and steady, and the opening not too small, so that a section of the spinal cord, for example, can be examined *in toto*, under a low power and with the diaphragm removed. The contrivance for revolving the stand is, as a rule, superfluous.

The cylindrical diaphragms (the disk-diaphragms are less perfect) must, as a matter of course, be exactly centered, and readily exchangeable. Narrow diaphragms are employed in the case of unstained objects, in order to bring the structures sharply into view when higher powers are used; with low powers it is usually necessary to attach a wider diaphragm, so that the entire field of vision may be utilized. A condenser, or Abbé's apparatus, is desirable in all cases, and it is necessary in examinations of schizomycetes. The rays of light, reflected from the mirror upon the lens of the condenser, are so refracted by this that they meet in a single point (focus), and this point lies exactly at the position of the object. In this way the object receives an immense quantity of light, not only a collection of nearly parallel rays from below, as in the ordinary examination with a narrow diaphragm, but an entire cone, with the largest possible angle of divergence, at the apex of which lies the object. For this reason the delicate contours of the transparent object, so far as they depend upon differences of refraction, are almost entirely lost; as Koch expresses it, the contours are "extinguished." The stained portions of the specimen, which would otherwise have been partly or wholly concealed by the outlines of the unstained portions, appear so much the more distinct. Koch terms this "isolation of the stained image." Thus, by the examination with the open condenser, we very often succeed in recognizing as such deeply-stained micro-organisms, or other small colored bodies, which under the ordinary illumination are concealed by the image of the structure, and thus are only indistinct or even quite invisible. The angle of divergence of the light-cone amounts, when Abbé's condenser is used, to 120°; the condensers previously manufactured usually give a much smaller angle, and are consequently insufficient. Beneath the lens of the condenser is a disk, provided with diaphragms of different sizes which are interchangeable; by employing a narrow diaphragm you naturally obtain an illumination quite similar to that furnished by a narrow cylindrical one. In order to isolate the colored image, the diaphragms are removed entirely. The other complicated arrangements of Abbé's apparatus have not, up to the present time, been essential to my purposes. On the contrary, the use of illumination by means of the open condenser is of very great value for all stained preparations, while in many difficult examinations it is even indispensable. We have to thank Koch for the introduction of this method.* In choosing a stand, care must be taken that a well-made condenser, with a large focal angle, or an Abbé's apparatus, is, or at least can be, attached.

2. OBJECTIVES. WATER AND OIL IMMERSIONS.—As regards the choice of the lenses, the necessary objectives are:

(1.) Quite a weak one, having a focal distance of about thirty millimetres, which with the medium ocular gives a magnifying power of about twenty; this is for the general inspection of large sections, those, for example, from the brain and spinal cord, liver, and kidney, for the observation of trichinæ, etc.

(2.) A moderately weak lens, with a focal distance of about fifteen millimetres, such as will attain a magnifying power of about sixty or eighty.

(3.) A moderately strong one, with a focal distance of four millimetres, giving a power of three hundred.

(4.) A rather strong immersion-lens, having a focal distance of one and one half or two millimetres, for more delicate examinations.

The powers of eighty and three hundred will be most useful for our purposes.

The immersion systems,† which are employed in order to obtain very strong magnifying powers, demand for their management a certain degree of care and experience. As already stated, it is best for the

* R. Koch. "Untersuchungen für Aetiologie der Wundinfectionskrankheiten." Leipzig, 1878.

† The usual designation of the immersion systems, as $\frac{1}{12}$, $\frac{1}{18}$, refers to their equivalent focal distance, which, according to the English custom, is expressed in inches: $\frac{1}{12}$ corresponds to 2 millimetres, $\frac{1}{18}$ to 1·3 millimetres.

beginner not to use them at first. It is impossible, without entering into rather exact physical demonstrations, to explain the advantages of immersion. This will be enough for us: as the rays pass from the upper surface of the cover-glass into the air, and again from the air into the under surface of the objective-lens of the microscope, those rays alone remain unchanged which impinge upon these surfaces vertically; the oblique rays are changed in their course, and the more acute the angle at which they fall, the more they are refracted.

Now, if we designate as the angle of divergence of an objective that angle which is formed by the diverging rays proceeding from any point in the object to the extreme edge of the lens (which are reunited at a point in the image above the objective), then it is evident that this angle of divergence can not be increased above a certain degree with dry objectives, if the clearness of the image is to be preserved.* For the peripheral rays, which, because of their oblique passage from the glass to the air, and from the air to the glass, are twice deflected, appear to proceed from a different point than the more central rays; in addition to the spherical aberration of light, there is another element which prevents the employment of an angle of divergence as large as possible. This evil is essentially lessened if we interpose a layer of water between the cover-glass and lens (water-immersion), since the difference in refractive power between water and glass is much less than that between

---

* This disadvantage is avoided, at least in part, by the so-called "correction." Compare the following note.

air and glass; it can be almost entirely removed, however, if a fluid is interposed which has the same refractive power as glass (homogeneous or oil-immersion). Oil of cedar, or a mixture of fennel and castor oils, is used for this purpose;* a mixture of chloral hydrate and glycerin has been recently recommended. These oil-immersion lenses are the masterpieces of our opticians; Abbé, Zeiss, and Stephenson have acquired a great deal of credit by their introduction.

With the increase in the angle of divergence, not only is the clearness of the image augmented, but the power of differentiation—the so-called resolving power—of the instrument is increased to a degree not hitherto attainable.

The use of oil-immersions is not unattended with difficulties, even for a skilled microscopist, who is accustomed to neat and delicate work.

In the case of water-immersion lenses, one very soon learns to determine the size of the drop, which is placed, by means of a glass rod, upon the front lens of the system, or upon the glass which covers the object; further than this, one has only to prevent

* This is only one of the advantages offered by the immersion systems; besides this they have, by reason of the relations of reflection and refraction, more light than dry systems of the same focal angle, as can readily be demonstrated. In order to correct the influence of cover-glasses of different thicknesses, the powerful dry and water immersion objectives are constructed in so-called "correction-mountings," which allow the lenses of which the system is composed to be approximated to, or separated from, one another. For cover-glasses of all thicknesses that position of the correction-screw must be determined at which the clearest microscopical image is obtained. The correction is, as a matter of course, superfluous with the homogeneous immersion.

the drop of distilled water upon the cover-glass from flowing over the edge, and mingling with the fluid in which the preparation is mounted.

This difficulty is absent in the case of cemented mounted specimens; the drop is then very easily removed from the cover-glass, by means of a fine capillary glass tube. An oil-drop, on the other hand, can only be completely removed by vigorous rubbing. We may generally content ourselves with getting rid of the superfluous oil by drawing fine blotting-paper gently over it, and may leave the very thin layer which remains upon the cover-glass.

3. EYE-PIECES. ACCESSORY APPARATUS. COMBINATIONS.—Two eye-pieces are generally used, a weaker one for ordinary work, and a stronger for special cases; one of these contains the micrometer. Of the accompanying pieces of apparatus, which are furnished by the optician, the following may be mentioned:

(1.) A revolving apparatus for rapidly changing the objectives.

(2.) An arrangement for drawing, preferably Oberhäuser's model, which has a knee-shaped bend, and is provided with two prisms.

(3.) A polarizing apparatus.

(4.) A spectroscope.

The last two pieces of apparatus have been recently combined by Messrs. Schmidt and Haensch of Berlin.

(5.) A warm stage—Schultze's or Stricker's.*

* There follows a list of German opticians whose instruments are recommended; this has been omitted as being of no practical use to the English reader.—TRANS.

## II.

## ACCESSORIES.

1. ILLUMINATING LAMP. COBBLER'S GLOBE.—Daylight always serves best for illumination, especially if it can be obtained from a white cloud which is somewhere in the neighborhood of the sun. Direct sunlight can not be used, therefore it is always most advantageous to place the working-table at a window which faces the south. If the sun is shining, a window is darkened by means of a white curtain, under the shade of which the work is carried on; the light is derived either from the curtain, or from the sky through the uncovered windows.

However, our clime, which so abounds in cloudy and dark days, frequently compels us, even during the day, to have recourse to artificial sources of light, especially for high powers. We use for this purpose merely a gas-flame, with an Argand burner, over which is placed an isinglass chimney, and a shade made of stiff paper. The direct light of the flame is used, and its yellow color is corrected by adjusting over the eye-piece a ring containing plane blue glass.

When we have at hand several of these rings, with varying shades of blue, we can make the correction for the different distances at which the lamp is placed.

The burner is fastened to a stand by a sliding arm, and stands at a height of about twenty or thirty centimetres above the surface of the table; if the shade is properly attached, by means of a wire frame, the heat of the flame need be felt little, if any; and then work by gaslight is not much more trying than by daylight. Petroleum-light is certainly useful, but one should see that the Argand burner has a diameter of at least twenty millimetres; the so-called duplex burners of corresponding breadth are also valuable.

I recommend highly the use of a cobbler's globe (Schusterkugel), which is filled with a solution of sulphate of copper and ammonia, and interposed between the lamp and the mirror. A few drops of ammonia are added to a solution of sulphate of copper, till a beautiful blue color results; by further diluting this solution with water, a cloudiness generally appears, which is redissolved by adding more ammonia. It is very easy to test the proper intensity of the color. A beautiful white light is then obtained, which falls upon the mirror in parallel rays.

The light itself must be placed low, and is only to be used for the actual microscopical work; for preparing specimens, etc., some other illumination must be employed.

2. GLASS APPARATUS.—The slides are to be made of white, plane glass, as free from flaws as possible, and of a definite, even thickness and size; the English are the best. The edges must be ground off.

The cover-glasses are likewise of a definite, moderate thickness, preferably about 0·15 millimetres. Many strong objectives require cover-glasses of special thinness. Among other glass objects the following are used:

A large number of watch-glasses.

Glass rods.

Glass tubes, capillary and otherwise.

Glass flat-bottomed dishes of different sizes.

Bell-glasses of different sizes.

Glass beakers, injecting-bottles, crucibles, reagent-glasses, and a stand with funnels.

Measuring-glass.

A plate of black glass and one of white porcelain, to serve as supports while preparing specimens. The black plate is adapted for white or unstained objects, while we always carry on our manipulations with stained preparations upon the white ground.

3. METALLIC INSTRUMENTS.—As regards the metallic instruments, needles, forceps, scissors, knives, spatula, etc., attention can not be directed too strongly to the fact that these must always be kept in an absolutely perfect condition. Even if we do see many older microscopists working with blunt, rusty needles, with dull scissors, with forceps which do not hold, etc., let this only serve as an example to deter us from a similar procedure. Our instrumentarium must be kept bright and sharp, like an oculist's case. The razor has its under surface ground flat; here, too, it is self-evident that the blade must always be sharp and clean. In spite of our double knife and microtome, we still need the razor very often, especially in hasty investigations. In cutting, *draw* rather than *press*, and utilize the whole length of the cut from beginning to end.

Within a few weeks nearly every one acquires the requisite experience, so that he can cut with the razor, rapidly and surely, a few even, thin sections, from fresh, as well as from hardened, preparations; in many cases, where absolutely exact or very thin sections are not required, this method, because of its simplicity, is always the best.

The double knife, also, is much used, especially for the examination of fresh preparations; with it large and regular sections are obtained, even from fresh, soft organs. However, the entire gross specimen is usually sacrificed in this way; the organs are frequently hacked to pieces with the double knife in a very ugly manner. The instrument is drawn quickly through the organ which is to be cut; in order that the blade may not strike anything at the same time and be injured, it is advisable to use a soft support, as fresh liver.

Most instrument-makers make the springs between the two blades of the double-knife too strong; I have often found it advantageous to remove these springs altogether.

The blades are thus adjusted: the upper screw is first screwed entirely back, then the two blades are closely approximated by the lower screw, and finally they are again separated a little, by advancing the upper screw. The blades should be nearly parallel.

4. MICROTOME.—No microscopist will be willing to work, at the present day, without a microtome. Although ten years ago microtomes were only in the hands of a few, now they are very widely used, and

it is certain that with them a most important advance has been made in microscopical technology.

It would be extremely tiresome (and, indeed, it is unnecessary) to introduce here a description of the various forms of microtomes, since several new models or modifications are annually invented. We can use almost any system; however, I would advise against those models in which it is necessary to imbed the preparation. Good microtomes are furnished by Dr. Long, of Breslau, Katschs, instrument-maker in Munich, and by the mechanicians Schanze, in Leipzig (at the Pathological Institute), Jung, in Heidelberg, and Meier, in Strasburg, and also by several others. The author worked for a long time with Long's instrument; but he is so well satisfied with Schanze's new model, that he would especially recommend this one. The apparatus explains itself: the knife is carried on a slide, while the specimen is gradually raised by the coarse movement of a toothed wheel; by drawing forward the knife, a section will be cut corresponding in thickness to the amount of elevation of the specimen.

In Long's microtomes, the specimen is raised by the limited movement of a slide upon a sloping plane; in Jung's instruments it is effected in a very accurate manner by means of a fine micrometer-screw. In these instruments, also, the sliding movement is remarkably smooth.

The specimen which is to be cut is secured in a clamp; clamps of different sizes and shapes, such as can be readily changed, may be used with the same instrument. In every case, the essential thing is that the specimen should be tightly held in the clamp.

It is customary to secure the preparation between two slices of well-hardened liver; amyloid liver, which has been kept in alcohol, is best fitted for this purpose. The thoroughly hardened specimen—a firm, even consistence is a necessary prerequisite for the making of good microtome sections—is placed in the clamp, between the two slices of liver, and is secured, by means of the screw, in such a way that a layer from one to two millimetres in thickness projects above the clamp. The slices of liver act like fixation-splints, by which the portion of the specimen which extends above the clamp is sufficiently secured. The fixation must always be perfect; if the specimen can slip away from the cutting-blade only a little, the sections are imperfect and unreliable.

We may, instead of this, fasten in the clamp a cork, upon the upper surface of which a slice of the hardened preparation, several millimetres in thickness, has been glued. As a glue, a thick solution of gum or so-called fluid cement is employed. Both substances soon become perfectly solid when placed in alcohol. In this way only a small piece of the specimen is used, and this is subject to no pressure, since it is merely attached to the cork. This procedure, which, as far as I know, originated with Weigert, is a very efficient one, and is to be highly commended.

When the specimen has been properly secured, the knife is so adjusted that it commences to cut at the point where the edge begins. This direction is, indeed, self-evident, but it is very frequently disregarded; the novice often pushes the blunt part of the blade against the specimen, or, on the other hand, he allows the edge to begin cutting at its middle. Both of these errors must be avoided. Then care must be taken that the knife meets the specimen at the most favorable angle. It should be pressed down as little as possible, but should have mainly a drawing movement; it is, therefore, to be so adjusted, with reference to the breadth of the specimen, that the entire edge is used up to its very end. The narrower the object, the more oblique is the knife, and the more economically can the section be made. In cutting, the specimen and the knife-blade are always kept thoroughly moistened with alcohol; the slide is to be well oiled, preferably with bone-oil.

Instead of the clamp, which is designed for hardened specimens, a freezing-plate or freezing-box may be attached to the same microtome; an ether-spray is thrown against the under surface of this, by means of an atomizer, an intense cold being produced by its evaporation.

A slice of any fresh organ, when placed upon the upper surface of the plate, freezes at once; in this way very delicate and even sections can be obtained directly from fresh organs, without previous hardening.

THICK AND THIN SECTIONS.—The thickness of the sections can be varied at pleasure, according as the specimen is more or less elevated; this thickness can generally be estimated directly from the micrometer-scale in hundredths of a millimetre.

The beginner generally thinks that in all cases the

thinnest possible sections must be prepared, while a practiced investigator often purposely works with quite thick sections. Very thin sections have the following disadvantages:

1. They are difficult to manipulate; it often takes a good deal of time to spread them out perfectly.

2. It frequently happens that the elements, inclosed within the meshes of extremely thin sections, fall out; this is highly disadvantageous for our purposes, since the very objects which interest us are often lost. Accordingly the process of penciling, or shaking, specimens, which is often practiced in normal histology, by which thick sections are rendered transparent, is to be recommended only to a limited extent in pathological examinations.

3. If we are searching for elements which are only sparingly distributed through an organ—for example, animal or vegetable parasites—there is naturally a better chance of finding the same in thick sections, provided that the latter are sufficiently transparent, and that the elements sought for stand out well from the surrounding tissue.

4. In thick sections definite stereometric representations of the structure of the object in question are frequently obtained, since a number of sections placed one above the other are examined directly *in situ* and *in continuo*, while in extremely thin sections plane images only are formed. On the contrary, it is evident that some particularly delicate structures generally become visible only in thin sections, while in thicker ones, in consequence of the many superimposed outlines, they completely disappear.

For most purposes sections of fresh organs having a thickness of 0·05 mm., or even 0·1 mm., are very useful; in the case of hardened preparations thinner ones are generally used, especially if they are stained, about 0·01 to 0·03 mm. being the average.

It is a very important advantage of microtomes that with them sections of any desired thickness can be produced easily and in any number, as if from a manufactory. More sections are usually made than are needed at the time; the rest are preserved for future use, in a small bottle which is filled with alcohol.

FURTHER TREATMENT OF SECTIONS.—The sections are transferred from the microtome-blade (or from the razor or double knife) to a watch-glass filled with fluid, and this is best done by means of a soft, moist brush. The fluid used for fresh (frozen) sections is salt-solution, and in this they are examined directly; for sections made from alcoholic specimens, first alcohol, and then, as a rule, distilled water are used. The sections may easily be ruined during the further manipulations to which they are subjected; to allow of the action of reagents and staining-materials they must frequently be transferred from one watch-glass to another, and finally be placed unharmed upon the slide. Slightly bent, thin spatulas are best used for this purpose—for example, strips of copper or pewter, or even of nickel-plated steel, which are brought carefully under the object as it floats in the fluid; in this way alone can a thin section be lifted out without wrinkling and be placed in a new fluid.

The slide also must first be covered with a thick

layer of fluid, before the section is transferred to it; the section should slip easily from the spatula, simply through the action of a gentle current of fluid. The cover-glass is then placed upon it, and the excess of liquid is absorbed by means of a capillary tube, or by blotting-paper. The slide rests meanwhile upon a black glass plate, in the case of unstained objects, or upon a white plate, when the objects are stained. The microscopist should also keep both hands free; it is a very bad habit with many beginners to hold the slide in the hand while placing the specimen upon it.

In order to render possible the delicate movements of the fingers, which are necessary for these and other manipulations, I advise that the fore-arm, and also the ulnar edge of the metacarpus, should be firmly supported upon the table; it is very difficult to mount thin sections properly when the arm is unsupported.

The sections prepared from alcoholic specimens are transferred from the watch-glass containing alcohol to distilled water; by reason of the active movements which are caused by the currents of diffusion, they spread out very nicely here, and become changed from their shrunken, wrinkled condition into transparent laminæ. Then for the first time do they become fit for examination; they are arranged upon the slide in distilled water, which is then in most cases displaced by glycerin, that has been placed at the edge of the cover-glass. Large, delicate sections do not unroll so well in the viscid glycerin as in water; smaller sections may also be placed at once in a drop of glycerin.

III.

REAGENTS. MICRO-CHEMISTRY.

Reagents are to be kept in small glass bottles with ground-glass stoppers. While this principle is carried out in every chemical laboratory, and even in every pharmacy, we constantly see many microscopists working with bottles which have dirty corks—a most reprehensible custom. Even a double stopper is proper for those reagents which are constantly used; we always keep our glycerin, acetic acid, distilled water, oil of cloves, Canada balsam, etc., in so-called cobalt-flasks, the ground-glass stoppers of which are drawn out below into a glass tube, while a hat-shaped glass cover is placed over the whole. The beginner should, by preserving his reagents in an absolutely clean and transparent condition, become accustomed, from the very first, to the most painstaking care in his work.

We, as a matter of course, alway employ preparations which are chemically pure; as yet we have never used any substances, except certain dyes, which can not be purified chemically. In order to insure cleanliness, besides the method of preserving reagents just mentioned, it is furthermore desirable that we should not use them in an uncleanly manner; he who dips a teasing-needle, brush, or even his finger, into his reagents, in order to remove a drop, will never be fit to undertake the more delicate investigations, especially the examination of schizomycetes.

Only a carefully-cleansed glass rod, or a glass tube which has been freshly heated, should be brought into contact with the reagent.

"ARTIFICIAL PRODUCTS."—The use of reagents is of the highest importance in histological examinations; many structural elements can only be studied by the aid of chemical influences. We, of course, endeavor to examine objects when they are as little changed as possible, in their normal—if possible in their living—condition; however, it would be a great mistake to reject as "artificial products" all the structures which can only be rendered visible by the use of certain reagents. No signs of a nucleus can be seen, as a rule, in the living white blood-corpuscle, and no cells in the living cornea, because the differences in the optical behavior of the nucleus and the protoplasm, and of the cell and the basement-substance, are too small to become visible, or rather, because the surrounding substance is too opaque to allow the delicate outlines of the inclosed body to appear through them. When death ensues, on account of various chemical changes, as coagulation, etc., these differences become more marked—that is, the enveloping substance becomes more transparent, and we are, at all events, justified in assuming that the nucleus in the living white blood-corpuscle, and the cells in the living cornea, are already present, although we are not able to demonstrate them until after *post-*

*mortem* change has begun; this is precisely the case with the borders of many epithelial cells, with the axis-cylinders of nerve-fibres, etc., which are also invisible in the living state, but are none the less present. Besides this, in examinations in pathological anatomy, we almost never observe the tissue in its living, normal condition, but always as the seat of more or less advanced cadaveric change. We must, accordingly, always bear in mind that a structure which is found regularly was not necessarily present as such *intra vitam;* however, it does point in every case to a differentiation which existed during life, and which is brought out by the cadaveric change, or by the reagent used.

*Furthermore, if among a certain number of elements, which originally appeared to be identical, some conduct themselves in a peculiar way toward a certain reagent, while others do not; if, for example, some are stained by a certain dye, while others remain colorless, we must necessarily conclude that there was a primitive difference between the elements.*

Upon this principle are founded all of the methods of preparation, some of them very complicated, which are employed for the exhibition of the different histological elements. From this simple observation it is evident what attitude we have to assume toward the microscopical images that are produced by our reagents.

We must not be confused by the expression "artificial product"; in former times many important histological discoveries were at first discredited in this way.

*In our microscopical investigations, we do not, as a rule, act simply as observers, but we experiment, and our results are consequently made up of preformed objects on the one hand, and of factors introduced by ourselves on the other.*

He would certainly be ever liable to the grossest errors, who should decide upon the question of original structure simply from his own results, without regard to these relations, which are generally very simple; an apparent fiber, for example, may either correspond to a genuine pre-existent fiber, or may represent a fold, or a production of coagulation, etc.

In our pathological examinations especially, we employ reagents for still another purpose. We often have the task of searching for certain elements, foreign bodies, parasites, etc.; if, now, we know that these elements which are sought for resist certain reagents and methods of treatment, while other substances are destroyed by this same treatment, then we have a very useful method of examination for our definite purpose, although the structural relations of the organ are entirely lost.

It follows from all this that microscopy, especially if it is concerned with pathological objects, should not be regarded as a purely mechanical process, for it frequently requires a certain amount of care and circumspection, even in the choice of the *modus procedendi*.

MICRO-CHEMICAL EXAMINATIONS.—Micro-chemical examinations are so conducted that either the specimen remains for some time in contact with the reagent (for example, it is placed in a watch-glass filled with the reagent, and is then examined under the microscope), or in such a way that the reagent acts upon the specimen while the latter is actually under the microscope. With this object, the reagent is placed at the edge of the cover-glass, and gradually makes its way toward the specimen; we can hasten this process by absorbing the fluid with blotting-paper at the opposite edge of the cover-glass. In this way the influence of the reagent may be directly observed under the microscope—for instance, the solution of granular protoplasm, of the red blood-corpuscles, and of lime, under the action of acids, etc. The beginner must naturally be extremely careful that the reagent does not get upon the upper surface of the cover-glass; in that case the object-lens of the microscope might very easily be injured. Other complicated reactions, especially most of the staining processes, are usually undertaken in watch-glasses; after having been acted upon by the reagent the specimen is compared with a model, or the same object is sketched first before, and then after, it has been treated.

The reagents principally employed are the following:

1. DISTILLED WATER.—Distilled water, as a rule, still contains small quantities of dissolved substances, and it furnishes (especially in summer) a favorable nidus for various minute organisms. If, therefore, micro-organisms are found in a specimen which has been treated with distilled water, this error must be guarded against; they can easily be removed by repeated boiling. The water diffuses at once, and very

actively, into most portions of the fresh tissues; in every case the vital properties of those elements of the human body which have been isolated in distilled water very soon vanish. The dead cellular elements, removed from the cadaver, are likewise essentially altered; the most rapid change takes place in the red blood-corpuscles; they swell, discharge their coloring-matter, and soon become perfectly invisible.

The method of employing distilled water in the examination of fresh tissues, and its limits are briefly as follows: We use it by preference in those cases in which we desire for our purpose to remove quickly from substances very rich in blood the blood-corpuscles, which, on account of their large number, frequently trouble us by concealing the other elements; however, we must never forget that the tissue itself may at length be essentially changed by the water.

If we are working with alcoholic specimens, distilled water simply causes them to swell, and generally in a symmetrical manner, so that nearly the original gross relations are restored. It is only seldom that further changes are induced in this instance, since the essential parts of the tissues, the albuminous bodies, are coagulated—that is, they are transformed into a variety that is insoluble in distilled water. However, we must always keep in view the fact that the diffusible substances that are soluble in water, as glycogen and sugar, are quickly removed from the sections.

2. SALT-SOLUTION OF 0·8 PER CENT. INDIFFER-

ENT FLUID.—The most varied micro-organisms very soon develop in this in great numbers; the fluids must therefore be very often renewed. I strongly advise against adding antimycotic substances, since we then no longer have a pure salt-solution; on the other hand, the solution may easily be sterilized by boiling.

In order to preserve the protoplasm and the red blood-corpuscles as nearly intact as possible, a salt-solution of 0·8 per cent is employed, the usual fluid in pathological examinations, which is suitable for sections of fresh tissues, as well as for the dilution of liquids. If we are particularly interested in preserving the vital properties of the cells for a longer period, we add to nine parts of the salt-solution one part of egg-albumen—the so-called artificial serum—or we use aqueous humor, hydrocele-fluid, transudations, blood-serum, etc.

3. ABSOLUTE ALCOHOL. HARDENING.—Always use the purest alcohol possible, and never the ordinary spirit, which, in addition to the water that it contains, is always contaminated with other substances, and frequently even has an acid reaction. In case we desire to use dilute alcohol, the absolute alcohol is mixed with the necessary amount of distilled water.

Alcohol diluted with two parts of water is sometimes used as an agent for the isolation of tissue-elements (Ranvier); while the cells become capable of resistance when placed in it, the cementing substances remain soft, so that the isolation of cells, which seemed previously to be fused together and to be attached to the intercellular substance, is an easy matter. To this end the bits of fresh tissue are immersed for about twenty-four hours in thirty-three per cent. alcohol.

The principal use of alcohol, however, is in the hardening of tissues, so as to give them a proper consistence for cutting. The process of hardening depends essentially upon two factors—the abstraction of water, and the coagulation of albuminates; beside this, alcohol removes from the pieces of the organ certain extractives, which are of no morphological importance, and a small amount of fat.

The immediate inference from this is that in all cases we can form a correct conclusion regarding the amount of fat in tissues (in pathological fatty degenerations, for instance) only by the examination of *fresh* organs, and *never* in alcoholic preparations.

Through the abstraction of water there naturally occurs a diminution, or shrinking of the specimens; if the different portions of the same object contain unequal amounts of water, the shrinking will take place in an uneven manner, and the specimen will be deformed in a very undesirable way. However, the form is generally quite well preserved, and corresponds to the diminution of the preparation due to the action of the alcohol; the sections swell up again if they are immersed in distilled water (through the absorption of water), so that they then resemble very closely their original condition. The main difference consists in the opacity of the sections, occasioned by the granular, coagulated albuminates, and this remains even after they have become swollen in water; as a remedy for this evil we generally use glycerin (q. v.) as an optical clearing-agent, or even acids and alkalies, by which the precipitated albuminates are redissolved, although many structures are destroyed at the same time. Hardening in alcohol is best effected by placing small fragments of organs in large quantities of absolute alcohol; in this way it is evident that the piece is entirely surrounded by the fluid.

A piece two or three cubic centimetres in size can thus be thoroughly hardened within twenty-four hours, and smaller bits still more rapidly.

The method, formerly in frequent use, of immersing the organs first in weak alcohol, and then in that of gradually increasing strength, has properly been altogether abandoned.

Alcohol is the best hardening-fluid for most tissues;* we employ it almost exclusively for this purpose. In the case of morbid specimens, it is of especial importance to us that the changes in their substance, occasioned by the method of preparation, should be simple and such as can be readily controlled; this is the action of alcohol, while the process of hardening in chromates, formerly so popular, which varies according to the differences in time, temperature, etc., causes changes, as clouding and staining, which it is very difficult to obviate.

A few little devices, which sometimes become necessary, are almost self-evident; for example, the eye is apt to shrivel up very quickly when placed in

---

* As has already been mentioned, hardening in alcohol is not adapted for examinations of fatty tissues.

alcohol. This condition can easily be remedied by injecting the fluid into the vitreous humor with a hypodermic syringe, during the early part of the hardening process, until the firmness and roundness of the globe are restored. This process must subsequently be repeated.

Many tissues, such as lung, muscle, etc., do not acquire a proper consistence for cutting, even after a long stay in alcohol;\* it is advisable in these cases to immerse the imperfectly hardened specimen for twenty-four hours in thin mucilage—equal parts of mucilage and glycerin. If the specimen saturated with solution of gum is again placed in alcohol, it hardens very evenly and firmly, since the gum is precipitated by the agent. The gum is very soon dissolved out of the sections when they are immersed in water. For the central nervous system alone, at least for the white substance, alcohol is not well adapted. In this instance its hardening effect is imperfect, corresponding to the smaller amount of water; besides this, alcohol abstracts a large part of the fatty matters of the nerve-medulla, which are then precipitated in a crystalline form, so that the tissue is greatly injured.

I have never been able to dispense with the chromates in the treatment of these important organs.

4. ETHER. CHLOROFORM. THE REMOVAL OF FAT.— Both substances are frequently employed in order to remove fats. As a matter of course they do not act

---

\* This is sometimes the case with rather old alcoholic preparations which are intended to be further hardened.

upon fresh tissues, since these are saturated with water, and chloroform and ether do not mix with water. The pieces (or sections) of an organ must first be dehydrated by treating them for some time with alcohol. The section from which fat is to be removed is then placed for about five minutes in a watch-glass filled with absolute alcohol, and then in a watch-glass containing ether or chloroform; if the fluid becomes clouded, it is a sign that the section is not yet sufficiently dehydrated, and it must be again placed in absolute alcohol. When it has remained for a few minutes in ether or chloroform,* so that the substances soluble in it are entirely removed, the section is transferred to alcohol for some time, and then to a watch-glass filled with water. It is examined either in water, or, as a rule, on account of the extreme clouding, due to the coagulation of the albuminates, acetic acid must be added in order to redissolve the albuminates. The order which is followed in the process of removing fat from fresh tissues is this:

Salt-solution or water.
Alcohol.
Chloroform or ether.
Alcohol.
Water to which acetic acid is added.

5. ACIDS. (*a*) *Sulphuric, Hydrochloric, and Nitric Acids; Decalcification.* — The strong mineral acids, when highly concentrated, possess the prop-

---

* In chloroform the sections become at once quite transparent; this does not depend upon the solution of the fats, but is a simple physical phenomenon, due to the high index of refraction of chloroform. If returned to alcohol the section immediately shows its former opacity.

erty of rapidly coagulating albuminous bodies; they can, in consequence, be used with advantage for fixing certain very delicate structures, as the so-called nuclear figures. According to Altmann,* it is best to immerse the pieces of organ for a short time (about one second) in a three-per-cent. solution of nitric acid, having a specific gravity of 1020, then to wash them in distilled water, and to harden in absolute alcohol. Flemming and other authors use a still stronger solution of nitric acid for the same purpose. When considerably diluted (about 1 to 1,000), the mineral acids cause essentially swelling of those substances which contain the most protoplasm (such as contractile or gelatinous tissue), the same as acetic acid.

The same reagents are used also to remove lime-salts. In order to make good sections through calcareous parts, such as bones, teeth, calcified tumors, etc., the lime must be removed; we seldom employ the earlier method of examining thinly-ground sections. This removal is accomplished most quickly by means of dilute hydrochloric or nitric acid. A one-half per cent. solution of the acid is used, and this is mixed with alcohol containing sodium chloride, in order to avoid the swelling of the basement-substance. The formula is as follows:

Hydrochloric acid . . 5 parts.
Sodium chloride . . . 5 "
Distilled water . . . 200 "
Alcohol . . . . . 1,000 "

The decalcifying fluid must be frequently changed, as it then gives quite brilliant results.

* Altmann, "Arch. für Anat. und Physiol.," 1881. S. 219.

A solution of chromic acid (about one per cent), or a saturated solution of picric acid, acts rather more slowly (Ranvier).

Moreover, we often find in a microscopical preparation deposits darker than the surrounding tissues, which lead us to suspect that we have to do with lime. We are convinced that they are lime if we observe that their dark outline disappears on adding an acid; in most cases the lime is combined with carbonic acid, so that on the addition of acid there occurs an active development of gas, presenting under the microscope a very elegant and striking picture.

If sulphuric acid is used, sulphate of lime, or gypsum, is formed, which is soluble with difficulty and crystallizes rapidly, in beautiful prisms, often uniting to form tufts. Gypsum-crystals are exquisitely double-refracting, as we can prove at once by means of a Nicol's prism attached in front of the eye-piece.

Many cementing-substances are soluble in strong acids; for example, a twenty-per-cent. solution of hydrochloric acid, or a thirty-three-per-cent. solution of liquor potassæ, is used to isolate smooth muscle-fibres. In order to isolate the urinary tubules over a wide area, so as to demonstrate their complicated course, strong hydrochloric acid is employed, with the application of gentle heat; however, this method has not yet come into use in the study of nephritis.

(*b*) *Acetic Acid.*—Organic acids, especially acetic, are very often used in our work; they are principally employed to dissolve, or to cause to swell, the albuminates and the gelatinous substance of which the connective-tissue fibrils are known to consist.

Since the substance of nuclei and elastic tissue, fats, the medulla of nerves, etc., resist acetic acid, this is a very convenient agent with which to expose the nucleus that is concealed within a darkly-granular cell, and the elastic tissue that is distributed throughout the connective tissue, in the substance of muscles. By the action of acetic acid also the fat-granules, deposited in protoplasm, in the contractile substance of muscles, etc., appear much more clearly. The micro-organisms which are present in the tissues behave in the same way.

Acetic acid acts in the manner described, even in a dilute solution of 1 to 100; in a solution of 1 to 1,000, also, its clearing action is still apparent, only it is somewhat slower.

If a section from a fresh organ, or from an alcoholic specimen, be placed in acetic acid, in a watchglass, it generally becomes quite transparent, and at the same time it swells considerably; this swelling usually takes place in an uneven manner, so that the section assumes a coarse, wavy character. It thus becomes nearly useless for examination. It is better to allow the acetic acid to act upon the sections beneath the cover-glass; a momentary raising of the glass, so as to allow the air to get under it, is sufficient to permit the drop of acetic acid, which has been placed at its edge, to make its way beneath, while the slight pressure of the cover is enough to preserve the even shape of the section. If the action of the acetic acid is to be rapid and energetic, the undiluted acid (glacial acetic) is used; for most purposes, however, it is advisable to dilute it with a little distilled water.

In many substances, which are saturated with alkaline albuminous fluids, acetic acid at first causes a clouding, due to the neutralization of the alkali; if more acid be added, this cloud clears up. However, a permanent opacity may be occasioned by acetic acid, which is not cleared up by excess of the acid; that is, mucin is precipitated by it. Fibrin, serum-albumin, and mucin are often found together in exudations, in the contents of cysts, etc. The action of acetic acid will vary according to the quantitative relations of the mixture; in most cases substances are rendered very clear and transparent by it.

With the marked swelling that acetic acid occasions in albuminous and gelatinous bodies, it is not surprising that the outlines which appear after the action of the acid, as well as the borders of the structures that resist it, do not always remain entirely unchanged. As far back as 1840 Henle called attention to the fact that the several nuclei which appear in white blood-corpuscles, pus-corpuscles, etc., after being treated with acetic acid, were not preformed, but that these contained only a single nucleus, which through the action of the acid was broken into several pieces. This actually occurs in many cases. Although we now know that some of the lymphoid cells, even in the living state, possess several nuclei, it is, nevertheless, highly important to remember that the nuclei that appear in a cell after it has been exposed to the influence of acetic acid may possibly represent artificial products, or fragments of a nucleus originally single.

In connective tissue acetic acid often causes an

essential change of structure; for instance, the regular rows in which the cells of tendons and fasciæ are arranged, are not usually to be distinguished after the swelling due to acids, while the nuclei are apparently strewn about in disorder. Ranvier took the precaution to fix upon a slide a small tendon, which was kept in a state of tension by means of little balls of wax applied to its ends, to cover it, and then to allow acetic acid to act upon it slowly; the swelling then occurs less irregularly, and the arrangement of the nuclei in rows becomes clear.

The beginner is strongly advised to convince himself of these and similar facts by personal experiment, in order that he may be in a position to decide what change he is inducing in structures by the addition of different fluids.

Formic and tartaric acids are less used; their action is similar to that of acetic acid.

(*c*) *Picric Acid.*—Picric acid, on the contrary, has a special use—namely, as a hardening and a staining material; the albuminates are gradually transformed in a saturated solution of picric acid into the insoluble variety, so that the tissues, with hardly any shrinking, assume a consistence proper for cutting.

Most substances are stained yellow at the same time, some very intensely, as smooth muscles, the horny cells of pavement epithelium, and of the epidermis, etc.

This characteristic staining also occurs very beautifully in sections which have been made from alcoholic specimens, and, indeed, in a very short time, within a few minutes; however, it is soon soaked out again by the action of water and alcohol. If we desire to preserve the staining, a small quantity of picric acid must be added to the water, alcohol, or glycerin.

(*d*) *Chromic Acid; Chromates; Müller's Fluid.*
—Chromic acid, when very dilute (about 1 to 10,000 or 1 to 20,000), is used as a macerating-fluid; if a small piece of spinal cord, for example, be placed in such a solution for twenty-four hours, it is then very easy to isolate the ganglion-cells with their numerous branching processes, while the cementing-substance is softened or dissolved. The action of chromic acid as a hardening agent is more important for our purposes; we use either a solution of the acid (having a strength of from one fifth to one per cent.), or else its salts. The principal ones in use are the bichromates of potassium or ammonium in a solution of about two per cent.; Müller also added sulphate of sodium to a solution of the bichromate of potassium. The following is the formula for Müller's fluid:

Bichromate of potassium . 2 parts.
Sulphate of sodium . . 1 part.
Distilled water . . . 100 parts.

It is much used as a hardening-fluid for the nervous system and the eye.

Perfect hardening occurs very slowly, only in the course of weeks and months, and the more slowly the larger the immersed pieces are, since the chromate penetrates gradually into the interior of specimens; six months or a year may be allowed for a cerebral hemisphere, but it has then a firm consistence. The hardening-fluid must be frequently changed, and in order to prevent the formation of mold in the solution a bit of camphor is added (Klebs). According to Weigert the hardening takes place much more rapidly if it is carried on in an incubator at a temperature of 30° to 40° C. Erlitzki suggests a fluid consisting of two and one half per cent. of bichromate of potassium and one half per cent. of sulphate of copper; in this preparation hardening occurs very quickly (in from eight to ten days) even at the temperature of the room.

The specimens are then placed in alcohol, which can be afterward slightly diluted.

The central nervous system after this treatment assumes a very even, firm consistence; at the same time it shows, even to the naked eye, certain characteristic differences in color. The gray substance is distinguished from the white by a brighter staining, while the latter becomes dark green; the usual form of gray degeneration, or sclerosis, in the white columns shows a dark brown shade, while most of the secondary degenerations take a brighter staining, even in those cases which did not display in the fresh state any difference in color between normal and degenerated portions. As regards other organs, hardening in solutions of chromic acid, or the chromates, formerly very popular, is to be recommended only in rare cases; except in the case of the nervous system and the eye, we much prefer hardening in alcohol, supplemented by the process of soaking in mucilage. Fibrous, or net-like, coagula are often produced by the chromates, and these may be erroneously regarded as preformed bodies; furthermore, the dark, granular precipitates within cells and interstitial tissues, due to their influence, are often quite troublesome, and are very difficult to clear up by means of chemical reagents. Deposits of lime are gradually dissolved by chromic acid and the bichromates, and may thus escape observation. Micro-chemical reaction can, as a rule, no longer be employed with specimens which have been treated with chromates; for these and other reasons, the latter are recommended as hardening agents only in those cases in which alcohol does not act favorably, on account of the peculiar chemical composition of the tissue—in fact, only in the nervous system, or in organs that are very fatty. It is always desirable to watch the results in freshly-examined or alcoholic specimens, on account of certain lime deposits which, through the action of chromic acid and its salts, may readily be completely dissolved and escape observation.

The simple chromate of ammonium, in a five-per-cent. solution, was used with great advantage by Heidenhain, especially in demonstrating the rod-like structure of the epithelium of the renal tubules. The agent is also to be recommended in pathological examinations.

6. ALKALIES. LIQUOR POTASSÆ AND LIQUOR SODÆ. AMMONIA.—Alkalies cause a breaking up, or swelling, of albuminates, of gelatinous material, of the contractile substance of smooth and striated muscles, and of nuclei; even horny structures are rendered perfectly transparent by these. About the only tissue-elements that resist are elastic tissue, fats

(including nerve-medulla), lime, pigment, etc., and amyloid material, beside chitin (the hooks of tæniæ, and echinococci), cellulose, threads of fungi, spores, and schizomycetes. From this it is at once evident how frequently we make use of alkalies; they come into play whenever we are looking for the bodies last mentioned. The structure of the tissue is, of course, almost entirely destroyed. While we can always gain a pretty good idea of a section that has been perfectly cleared by acetic acid, when alkalies are used everything disappears and we have no guides except the elastic tissue and the homogeneous membranes. Liquor potassæ or liquor sodæ (in a solution of from one to three per cent.) is best used for most purposes; the clearing action begins at once, even with this dilution. The concentrated lye (thirty-three per cent.) has a special action; in this solution most of the elements are preserved, while the cement-substance is dissolved. This applies particularly to smooth and striated muscular fibers. If, for example, a bit of a uterine leio-myoma is placed for a few minutes in a watch-glass filled with a thirty-three-per-cent. solution of liquor potassæ, it separates into its individual fiber-cells under the needle, almost of itself; only care must be taken that the lye does not become diluted, for in that case the fibers themselves at once dissolve. The preparation must be examined directly in the lye.

Even red blood-corpuscles preserve their shape in the thirty-three-per-cent. solution, while they at once disappear in dilute solutions.

According to an observation of Virchow, weak

alkaline solutions are able to excite the movements of ciliated epithelia after they have become motionless and apparently dead.

7. GLYCERIN.—Glycerin must, above all, be without a trace of acid; a small quantity of water does less harm. It is generally used in a pure form, since when diluted with water it is very apt to become moldy. Glycerin is of great value in the histological examination of organs which have been hardened in alcohol, and in other fluids that coagulate albumen, such as picric and chromic acids and their salts. During this process the tissues have necessarily become much clouded; if acids or alkalies are used, in order to dissolve the albuminous granules that have been precipitated by the hardening agent, many other structures are destroyed at the same time, such as connective-tissue fibers, fibrin, and blood-corpuscles.

For these cases, therefore, glycerin is used as a clearing agent. The clearing action of this fluid is due, not to the chemical solution or swelling of the albuminous granules (fat alone is gradually dissolved in glycerin), but rather to a physical force, to its high refracting power. We can at once demonstrate this action to ourselves by comparing the outline of a glass rod which is dipped in water with that of a similar rod dipped in glycerin; the latter is much more delicate. Or, if one piece of filter-paper is saturated with water, and another with glycerin, the latter becomes much more transparent. The outlines of a fragment of tissue which has been moistened with glycerin are altogether more delicate; glycerin is therefore useless for the examination of most fresh tissues, the elements of which already possess delicate outlines, since the latter then become almost invisible. On the contrary, the degree of transparency effected by glycerin is peculiarly adapted for alcoholic specimens. It may be said that since its introduction into microscopical technology the examination of organs hardened in alcohol has reached its full development. Glycerin mixes with water, also with alcohol, acetic acid, etc., in all proportions, but rather slowly because of its sirupy consistence, so that a glycerin-preparation is a suitable one for the induction of a rapid chemical reaction, for example, the action of iodine or an acid. However, it is very easy to remove glycerin from the preparation, by simply placing it in a watch-glass filled with water.

Glycerin has also the well-known property of neither evaporating when exposed to the air, nor undergoing any other chemical changes; at the most it only absorbs a little water, under some circumstances. This property renders it an excellent material for the preservation of microscopical specimens. If we desire to preserve a preparation which is in water, or a watery solution, it is only necessary to place a drop of glycerin upon the edge of the cover-glass; the glycerin flows under the cover as the water evaporates. Fresh preparations may also be kept in this way; if the glycerin is subsequently displaced by water or salt-solution, the original condition is restored.

The dark shading, and the shining appearance of

the elastic fibers and laminæ, are only slightly diminished in glycerin, since their refractive power is considerably higher than that of the latter; on the contrary, the characteristic luster presented by amyloid material, and by Recklinghausen's so-called hyalin, and other colloid bodies, when examined in watery fluids, disappears almost entirely in glycerin, since their refractive power differs only very slightly from that of the latter. On careful examination the difference is still somewhat apparent in most cases; it is always well to examine the specimens first in water, if you are looking for these objects. It has already been stated that small fat-drops disappear entirely in glycerin, hence glycerin-preparations should never be used in looking for fatty degeneration.

8. ACETATE OF POTASSIUM.—A saturated solution of acetate of potassium (as recommended by Max Schultze) may also be used as a preserving-fluid; it does not evaporate, and is stable in the air. It has only a feeble clearing power, hence it is to be employed especially for keeping fresh objects that have not been hardened. This method is quite useful for preserving fatty degenerated tissues; however, the outlines of the fat-drops and, in time, their original distinctness, is lost.

9. OIL OF CLOVES. CANADA BALSAM. — If we wish to clear preparations more thoroughly (especially after they have previously been stained intensely), we use oil of turpentine, or, what is to be recommended more highly, oil of cloves; other ethereal oils, such as oil of cedar, origanum, cinnamon,

bergamot, anise, etc., also xylol and phenol, act in a similar way.

Every one can select that substance which is least disagreeable to his olfactory organs. These fluids are either not at all, or only slightly, miscible with water, so that the sections that are to be cleared are first dehydrated with alcohol; it is enough to place them for a few minutes in a watch-glass containing alcohol, after which they at once become saturated with oil of cloves. The sections thus attain the highest degree of transparency; the refractive power of the fluids mentioned is very great, much greater than that of glycerin, and nearly the same as that of glass. The most resistant outlines of the elements of animal tissues disappear almost completely under this treatment; in unstained preparations nearly everything vanishes, as a rule, and even elastic fibers are difficult to recognize, especially if open illumination by means of the Abbé-Koch condenser is used; but the stained portions appear so much the more distinctly. With this method of examination, therefore, we must always keep in view the fact that most of the structures have been purposely excluded from our view. Specimens thus prepared can at once be permanently preserved in resinous mounting-materials; we generally use Canada balsam, or xylol, dissolved in equal parts of chloroform. A solution of mastic in chloroform is used, also dammar-varnish, etc.

The oil of cloves is absorbed by fine blotting-paper, and its place is gradually occupied by the Canada balsam, which is deposited at the edge of the cover-glass.

REAGENTS USED IN THE PROCESS OF STAINING. THE PRINCIPLES OF STAINING.—The art of staining has become every year more important and indispensable. Among the results obtained by staining the most striking was the discovery of vegetable parasites. Weigert, Ehrlich, and Koch have won the highest distinction in this field; the very valuable studies of Ehrlich related substantially to the theory of the action of dyes.

*The principle in staining is, therefore, that certain elements of tissues, and also of cells, appropriate actively, or in large quantity, from the solution employed, a certain dye, and form with this a combination having an intense color, that is more or less permanent.*

The relation of the different substances of the human body to the different dyes is naturally a highly complex one, and we have gradually learned to recognize the fact that for almost every tissue-element there is a special dye, or a special method of staining, by the action of which it takes a deep characteristic color, such as distinguishes it from other tissue-elements.

*Hence, in many instances staining assumes the importance of a chemical reaction, by means of which any particular structure, that lies concealed among other bodies, can be brought easily and conveniently into prominence.*

This "elective" action of dyes is of extreme importance in pathological investigations, as will at once be evident; elements which, because of their delicate contours, are only to be recognized after very careful examination, in the midst of the confused mass of other structures, after "isolated" staining, are seen at once, even at a cursory glance, and often with a low power, so that we are soon compensated for the time spent in the staining process by the gain in convenience and accuracy. The process of staining alone frequently discloses to us existing differences in tissue- and cell-elements which had previously appeared to be perfectly identical.

It has already been stated that the perfect application of the results of staining was first attained by means of the open condenser, or Abbé's illuminating-apparatus. The technique of dyeing is usually this: A section is transferred from distilled water to a dish filled with the staining-solution, with which it is entirely covered; it remains in this for different lengths of time, varying from a few minutes to twenty-four hours, and is again immersed in distilled water, in order to wash away the portions of the dye that are adherent to its exterior; it is then examined, either in glycerin directly, or in oil of cloves, after dehydration with alcohol. The *rationale* of the staining process in this case is simply that certain elements take the dye, while others remain unstained. This is called "election."

In many cases, however, the section which has been removed from the staining-solution and washed is subjected to further manipulations; it is again decolorized, that is partially. In this instance there has occurred at first a diffuse, even, but unnecessary amount of staining; but during the supplementary process of extraction, while certain elements give up their staining completely, others, that have a stronger affinity for the dye, retain it. This is called by Ehrlich the principle of maximum staining. This process, which indeed was first employed by the author, now plays a prominent part, especially in staining with aniline dyes; alcohol, or acids, serve as the ordinary extracting-agent.

It is only in exceptional cases that it is advisable to undertake the staining of a section under the cover-glass; the dye is uneven and is confined to the edges. Isolated elements, such as cells, can sometimes be stained under the cover; however, in this case also, the method of coloring dry preparations, employed by Koch and Ehrlich, is, as a rule, much to be preferred. (Compare the section on "The Examination of Fluids.")

It has long been the custom in embryological and zoölogical investigations to stain organs or animals *in toto;* alcoholic fluids, especially, have been compounded which effect simultaneously the hardening, as well as staining, of the preparations. The advantage is that the section can be transferred almost directly from the microtome to the slide for examination; aside from the shortness and convenience of the process, there is far less danger in this way of injuring or destroying the section during the different acts of staining, washing, etc. But I believe that this method is seldom adapted to our ends; in many, indeed in most cases, we have first the task of giving a definite opinion concerning a concrete, practical case, and we must therefore always leave ourselves the opportunity of applying every available test to the specimens under consideration. We should block our own path, as it were, if we carelessly adopted the method employed in normal anatomy, since the relations are totally different. The normal anatomist, for instance, examines as many eyes as he chooses in order to decide upon a scientific question; every normal eye is therefore of equal value to him, and for every variation in his method he can use a new specimen. But we are often limited to the single specimen before us, from which we must decide upon the structural changes that have occurred. To this end we must take care to obtain from the object (which has been changed as little as possible by freezing, hardening in alcohol, etc.) a series of sections, made through the diseased parts, which we may regard as somewhat similar specimens. We may then treat these sections according to different methods, and experiment with them, in order to study the changes from every side. We can never know what surprises we are about to encounter in the interior of diseased organs, so that a simple way of preparing objects, which disturbs them as little as possible before they have been dissected, is the best one for us. It is impossible for us before beginning the examination to adhere to a fixed method, but we must reserve for ourselves the possibility of taking the longer (since it is the necessary) road, according to the result of examination, with which we are not yet thoroughly acquainted.

We shall, therefore, not describe in this volume the process of staining entire organs; if any one wishes to employ it, he will obtain the necessary hints from Grenacher ("Archiv. für mic. Anat.,"

Bd. 16), and P. Myer ("Mittheilungen aus der zoöl. Station zu Neapel," Bd. 2, 1880).

10. IODINE.—This, the oldest of the staining-materials that have come into use in microscopic examinations, is still used very frequently in the form of Lugol's solution. Iodine, insoluble in pure water, is readily dissolved in a solution of potassium iodide; the following mixture is used:

    Pure iodine . . . . . 1 part.
    Iodide of potassium . . . 2 parts.
    Distilled water . . . . 50 "

This solution can be diluted to any desired extent, if necessary. It should be observed that it is difficult to preserve the iodine-staining in water and glycerin for any length of time; iodine always forms merely a loose combination with organic bodies, and gradually evaporates, when the staining vanishes. Even when the preparations are carefully mounted, it is apt to disappear in a few years at the latest. They can not be kept in Canada balsam at all, since the color is at once extracted by alcohol. On the other hand the iodine-staining seems to be held fast in specimens that are placed in a thick solution of gum. Albuminous, as well as gelatinous and colloid substances, are readily colored yellow by the iodine solution; the cells are usually more deeply stained than the interstitial tissue, and the nuclei rather more deeply than the protoplasm. The cellular elements (as the columns of cells in carcinoma) can accordingly be brought out in fresh sections by a very rapid and convenient staining process.

The red blood-corpuscles are stained dark brown

by iodine. The following substances give a peculiar reaction with the solution of iodine:
Glycogen,
Corpora amylacea,
Amyloid material.

*Glycogen*—In many cartilage-cells, as in those of the chorda dorsalis, in the proliferating layer of epiphyseal cartilages, even in the normal condition, but in an especially striking manner in rickets, as well as in enchondromata, an intense wine-red staining is obtained with the iodine-solution, which affects either the whole, or only a part, of the cell-substance. The reaction appears best when the rest of the substance is quite faintly stained. The red color depends upon the glycogen contained within the cells, as Neumann and Jaffé have shown; when not stained with iodine the portions that are rich in glycogen often present a homogeneous, glistening appearance. From a discovery made by Claude Bernard, the same condition was long ago recognized in the cells of the chorionic villi and other embryonic structures. Boch and Hofmann have also, by means of iodine-staining, studied the variations in the amount of glycogen contained within the liver-cells; Schiele, a pupil of Langhans, also discovered a large amount of glycogen in normal stratified pavement-epithelium as well as in rapidly-growing cancers.

Ehrlich recently found that if the sections remain in watery solutions (it is well known that glycogen is insoluble in alcohol) a large part of the glycogen is extracted from the cells, and in consequence escapes observation; he employs, therefore, in order to avoid its solution or diffusion, a dilute solution of iodine mixed with mucilage, in which the specimens are directly examined and preserved. By following this method Ehrlich ascertained that in diabetes large quantities of glycogen regularly appear in the epithelial cells of the renal tubules, especially at the boundary-line between the cortex and medulla.

*Corpora Amylacea.*—It is well known that starch-granules are stained an intense blue color by iodine; they are frequently found in the contents of the stomach and intestine, and in the cavity of the mouth, and are positively recognized by the iodine-reaction; they are often met with also as chance impurities. The so-called corpora amylacea, which are found almost invariably in the nervous system in degenerative processes, and in the white substance of the brain and cord in elderly subjects, bear a certain purely external resemblance to starch-granules; they are to be regarded as products of degeneration of the medullary sheath. They assume a deep wine-red color with iodine. Certain concretions also, which are found occasionally in the lungs, and very often in the prostate, are also described as corpora amylacea. All these transparent or yellowish-brown bodies are distinguished by their concentric lamination, and by the fact that they stain more or less intensely with iodine; the shade varies from wine-red to dark brown. As to their nature and significance little is known; they have nothing to do with starch, and still less with amyloid material.

*Amyloid Material.*—Amyloid substance is characterized by the wine-red color which it forms with iodine; it is better to use in this case also rather weak solutions, having the shade of cognac, so that the reaction may take place gradually, in the course of several minutes, and be on that account so much finer and clearer.

In many cases of amyloid degeneration the color changes to a dark green or blue tint on the addition of sulphuric acid; the sections that have been stained not too deeply, but only a clear yellow, are placed in a dish containing a one-per-cent. solution of the acid, when the reaction occurs at once, or in the course of a few minutes. As may be inferred, this blue or green staining—by which the degenerated portions are brought out from the surrounding tissue in a far more striking manner than when iodine-staining alone is used—does not occur in all cases of amyloid degeneration. The sulphuric acid frequently does not alter the shade at all, but merely produces a saturated brown color. Furthermore, in many cases of amyloid change we find that some of the degenerated elements become green or blue when iodine and sulphuric acid are added, while others only take a dark red staining; in several cases that came under my observation, for example, the arteries and vasa afferentia of the kidneys became dark red, while the loops of the glomeruli, on the contrary, were dark blue. This colored picture, which by reason of its regularity is extremely striking, is not very often seen; on the contrary, the amyloid arteries are found for the most part to be stained red, only a few isolated portions being colored blue. These differences in staining may possibly be related to the age of the amyloid material, the younger portions being colored red with iodine and sulphuric acid, while the older become blue; thus I have several times observed in the spleen that the arteries and capillaries of the enlarged follicles were stained blue, while the vessels of the less (or more recently) diseased pulp were stained red. Further differences between these two varieties of amyloid have not yet been established; when no staining is used, or when coloring with aniline dyes is employed, only a uniform appearance is seen, whereas the treatment with iodine and sulphuric acid brings out a striking difference in color. The iodine reaction has hitherto been indispensable for the microscopical diagnosis of amyloid; amyloid shares its homogeneous, glistening character with other colloid and hyaline substances, but the latter are only stained a faint yellow by iodine. The red color resulting from treatment with aniline dyes is not always characteristic, as it appears; for example, we frequently succeed in staining urinary casts red with aniline-violet, while they are only rendered yellow (or brown if the action is stronger) by iodine.* The inference is that even the casts that are stained red with aniline do not consist of real amyloid substance, but perhaps represent an initial stage in the formation of amyloid.

The chemical character of amyloid material has been much studied; it is known to be a body rich in nitrogen, and related to the albuminates. On the other hand, the cause of the iodine and iodine-sul-

* In certain instances casts are also stained reddish-brown with iodine.

phuric acid reactions, from which amyloid derives its highly inappropriate name, is still entirely unknown. Nothing is known concerning the red iodine combination and the blue bodies into which the latter is changed through the action of sulphuric acid.

It should be noted that cholesterin also stains quite dark in a dilute solution of iodine. If a drop of strong sulphuric acid is allowed to flow under the cover-glass, a beautiful blue color also appears at the angles of the tables.

11. CARMINE.—The introduction of carmine staining dates from the year 1858; we are indebted for it to Harting and Gerlach.

(*a*) *Ammonia-carmine.*—According to Gerlach's original formula, the carmine was dissolved in ammonia. Add to one part of finely powdered carmine one part of strong liquor ammoniæ, and from fifty to one hundred parts of water. The mixture is exposed to the air for twenty-four hours, in order to allow the greater part of the ammonia to evaporate, and is then filtered.

The less free ammonia there is in the solution, the less injurious is its action upon the tissues; the solution must be frequently renewed, however, since it is easily ruined by the formation of mold.

The carminate of ammonium stains very rapidly a large number of the substances found in animal bodies. The staining is genuine—that is, it is permanent—especially if the section, after careful washing, is placed in dilute acetic acid, for the purpose of fixing the color. If the preparation has not been washed in exactly the proper manner, it will be completely spoiled by the granular precipitate of carmine produced by the acid.

The following parts are stained: Protoplasm, the nuclei of nearly all cells, the fibrillated basement-substance of connective tissue, smooth and striated muscular fibers, the basement-substance of osteoid tissue and of decalcified bone, fibrin, the neuroglia of the central nervous system, the axis-cylinders of nerves, most colloid substances, etc.

The basement-substance of hyaline cartilage, elastic tissue, horn, the medullary sheaths of nerves, fat, mucus, calcified bone, etc., remain unstained.

Ammonia-carmine is used principally for examinations of the nervous system, in order to bring out the axis-cylinders.

Nerve-tissue, as before mentioned, is generally hardened in solutions of the chromates; the longer the stay in the latter, so much the slower and more difficult is the staining process apt to be, so that it often requires several days before the proper shade is attained. This drawback is quite annoying, especially in the case of very small axis-cylinders (in the optic nerve, for instance), that require the deepest staining. The staining can be rendered more rapid or intense if the solution is placed in a compartment warmed to about 50° C. (Obersteiner). Under these circumstances the color appears in a satisfactory manner, even within an hour.

Heule and Merckel devised another method which is worthy of high commendation. Place the section first in a solution of chloride of palladium (one to five hundred) for about ten minutes, when it is dyed straw-yellow; it is then transferred to the carmine-solution, in which it is stained red in a few minutes. After careful washing in water, dehydration in alcohol, and clearing in oil of cloves, it is examined. By this method the medullary sheaths are stained yellow, the neuroglia, ganglion-cells, and axis-cylinders a deep red. If it is desirable to bring out the nuclei also, they can be shown very beautifully in the carmine section by subsequent staining with hæmatoxylin (compare page 59). Such double stainings are to be highly recommended, especially for the examination of secondary degeneration of the spinal cord, sclerosis, tabes, etc.

The diffuse red color of the neuroglia in specimens thus prepared is often troublesome. In order to eliminate this, Ranvier places the sections, after they have been stained with carmine, in a mixture of one part of formic acid and two parts of alcohol, for from five to ten hours; the axis-cylinders and nuclei then remain red, while the neuroglia is deprived of color ("Comptes Rendus," November, 1883).

Aside from this, ammonia-carmine is generally used only for the examination of the osseous system. Here, too, double staining (with hæmatoxylin) acts very well. The bringing out of the osteoid tissue, by means of ammonia-carmine, is of real value, especially in the examination of rickets and osteomalacia, which is best undertaken with fresh specimens—that is, without artificial decalcification.

OTHER CARMINE DYES.—A large number of modifications of carmine-staining have been proposed and recommended; we mention here only a few of these, such as are especially valuable for pathological purposes.

(*b*) *Picrocarmine* (Schwarz, Ranvier).—The picrocarmine of the druggists is generally useless; according to Weigert, a small quantity of acetic acid is to be added to this in order to form a good staining-material; if a precipitate occurs, it is readily dissolved by a trace of ammonia.

The author prepares a picrocarmine that stains very rapidly according to the following formula: To one part of alkaline ammonia-carmine solution (consisting of one part each of carmine and ammonia, and fifty parts of water) add slowly, then drop by drop—while stirring constantly—from two to four parts of a saturated solution of picric acid, until the precipitate that first forms is no longer dissolved by stirring; the greater the quantity of ammonia, so much more picric acid is it necessary to add. The fluid is then filtered, a few drops of phenol are added to every one hundred centimetres, in order to preserve it, and a cloudiness that appears somewhat later is readily dissolved by adding a trace of ammonia.

This fluid is very useful for most purposes; it produces a double staining within a few minutes. The nuclei are all colored deep red, the fibrillated substance of the connective-tissue takes a faint reddish tinge, while the protoplasmic bodies, on the other hand, the smooth and striated muscle-fibers, horn, most hyaline and colloid substances, etc., take a more or less intense yellow.

The difference is often still more striking if the

sections after being stained are placed for half an hour in a dish filled with glycerin containing hydrochloric acid—one part of hydrochloric acid to one hundred parts of glycerin; the picrocarmine solution stains with especial rapidity and intensity if it contains a little free ammonia; in this case, however, the carmine dye first preponderates. By treating with the acid glycerin the red dye is first removed from the protoplasmic and interstitial substances, so that the yellow picrine-staining becomes prominent (Neumann); the red color, on the other hand, is fixed in the nuclei at the same time. It must be observed, moreover, that the red staining of the nuclei (in neutral or acid fluids) is permanent, while the yellow picrine dye is soon soaked out. In order to retain the yellow shade, it is customary to add a small amount of picric acid to the water, glycerin and alcohol employed, until a slight tinge of yellow appears. Then the preparations stained with picrocarmine can be mounted in glycerin just as well as in Canada balsam.

Picrocarmine is a very valuable agent on account of the convenient double staining that it causes, showing clearly the nuclei on the one hand, and the protoplasm of hyaline substance, of horn, smooth muscular fiber, etc., on the other. The discovery of tubercles in strumous granulations, and in the tissue of lupus, has been rendered much easier for the author by the employment of this method; picrocarmine is also to be highly recommended in examinations of the nervous and osseous systems, and of many glands.

(c) *Borax-carmine* (Grenacher).—Mix in a porcelain dish and heat to the boiling-point:

| | |
|---|---|
| Carmine . . . . | ½ part. |
| Borax . . . . . | 2 parts. |
| Distilled water . . . | 100 parts. |

Dilute acetic acid (about five per cent.) is to be added drop by drop to the bluish-red fluid, while stirring constantly, until the color changes to that of the alkaline ammonia-carmine solution; allow it to stand for twenty-four hours, decant and filter, a few drops of phenol being added as a preservative.

A section immersed in this solution is stained deeply in a very short time, even in a few minutes; however, the staining is quite diffuse, and is therefore useless. On the contrary, the most beautiful staining of separate nuclei is obtained, if the deeply stained section is placed in a dish full of alcohol that contains hydrochloric acid, prepared according to this formula:

| | |
|---|---|
| Hydrochloric acid . . . | 1 part. |
| Alcohol . . . . | 70 parts. |
| Distilled water . . . | 30 parts. |

The section at once discharges a part of the dye, and is surrounded by a red cloud; after an interval, varying from a few minutes to half an hour, it is washed (in water or in alcohol) and is examined in glycerin or in oil of cloves. This method gives the most intense nuclear staining; in making use of it the action of the hydrochloric acid must be taken into account—viz., the solution of lime, swelling of fibrin, protoplasm, fibrillated substance, etc.

(d) *Alum-carmine* (Grenacher).—A gramme of

carmine is heated with one hundred cubic centimetres of a five-per-cent. solution of alum; boil for twenty minutes, and filter after cooling.

A staining, confined almost exclusively to the nuclei, can be obtained with this solution within from five to ten minutes, but it is not quite so intense as in the former case.

(*e*) *Cochineal-Alum Solution* (Partsch and Czokor).—One part of the finest cochineal (the original substance of carmine) and one part of alum are heated in one hundred parts of water until about half of the solution has boiled away; a small amount of phenol is added and it is filtered.

The action is quite similar to that of the former solution; I prefer to use it for the simultaneous staining of nuclei and axis-cylinders in sections of the central nervous system, after previous hardening in the chromates. The staining takes place within twenty-four hours, and the nuclei have a different shade (deeper violet) than the axis-cylinders.

(*f*) *Lithium-carmine* (Orth). — Two and one-half parts of carmine are dissolved in one hundred parts of a saturated solution of carbonate of lithium. The sections are stained in a few minutes; after decolorizing in alcohol containing hydrochloric acid (comp. under *c*) a beautiful nuclear staining appears.

Picro-lithium-carmine is prepared by adding to the solution described from two to three parts of a saturated solution of picric acid. The color also appears rapidly, and presents the advantage of double staining, as in the case of picrocarmine. Here, too, some of the dye may be removed by means of alcohol or glycerin, mixed with hydrochloric acid, when the contrast becomes greater.

Since this solution is also quite permanent, it can be highly recommended.

12. HÆMATOXYLIN. *Weigert's Method of Staining the Central Nervous System.*—Staining with hæmatoxylin is one of the surest and most excellent ways of bringing out the nuclei clearly.

Hæmatoxylin crystals readily dissolve in alcohol, forming a brown tincture; if a small portion of this tincture be added to a watery solution of alum, a bluish fluid results after a few minutes, the staining power of which, however, does not reach its full height until several days have elapsed; at the same time, or soon after, the dye also begins to be precipitated in a granular form, so as to spoil the specimens. The solution must always be filtered, therefore, just before it is used. In order to obtain a permanent solution, having a constant staining power, the following formula is recommended:

Hæmatoxylin,
Alum, of each . . . 2 parts.
Alcohol,
Glycerin,
Distilled water, of each . . 100 parts.

A little acetic acid may afterward be added to the mixture, in order to prevent overstaining (Ehrlich). It should be observed that the solution only attains its full staining capacity eight days after its preparation.

A section immersed in this brown solution is stained a similar color in a very short time; the section is washed in distilled water, and in the course of a few minutes its color changes to blue. A staining confined almost entirely to the nuclei and to (most) schizomycetes is then observed. If other tissue-elements have been stained at the same time, these should be decolorized with alcohol containing hydrochloric acid; the staining of the nuclei is well preserved. For many purposes it is desirable to color the protoplasmic substance also; this can be accomplished by subsequent staining with picric acid (saturated solution), or with eosin (see under Eosin). The specimens stained in hæmatoxylin gradually lose their color when kept in glycerin, and hence are best preserved in Canada balsam.

Weigert* has taught us quite recently a most valuable use of hæmatoxylin staining, in connection with the central nervous system, whereby it is possible to exhibit, in a very elegant manner, the fine medullated nerve-fibers, which could previously be shown only with the greatest difficulty. The method is a peculiar one, and is somewhat complicated. The portions of the central nervous system are hardened in Müller's or Erlitzski's fluid (compare page 34), and are then transferred to alcohol without previous soaking; the sections, before they are dyed, must also be placed, not in water, but only in alcohol. The staining-solution consists of one part of hæmatoxylin, ten parts of alcohol, and ninety parts of water; the mixture is boiled, and allowed to stand for a few days. The sections stain best if they are placed for one or two hours in an incubator at a temperature of 40° C.,

\* "Fortschr. d. Med.," Bd. ii, S. 190.

and are then washed in water. The deep black sections, which are greatly over-stained, are then almost entirely decolorized by immersion for half an hour or an hour in a mixture of two and one half parts of ferrocyanide of potassium, two parts of borax, and one hundred parts of water, until the gray matter appears yellow; the white substance remains black. The sections are then thoroughly washed in water (ferrocyanide of potassium is precipitated by alcohol), and placed successively in alcohol, xylol and Canada balsam.

With this staining the nerve-medulla comes out very sharply, by reason of its dark color, while the axis-cylinders, ganglia, cells, nuclei, etc., remain nearly colorless.

By employing Weigert's method, I have succeeded in demonstrating positively the change in the gray substance in tabes, which has long been asserted, but which has hitherto been sought for in vain—namely, the disappearance of the network of delicate nerve-fibers in the interior of the columns of Clarke. This discovery is of great theoretical importance. I have described the method minutely, since it is quite possible that other changes will be found in the cerebral cortex in progressive paralysis, secondary degenerations, retinal affections, etc. The specimens obtained in this way are extremely useful; the nuclei may easily be stained subsequently, by means of alum-carmine.

13. EOSIN.—Eosin forms a fluorescent solution, giving a rose-red color with transmitted, and a greenish color with direct light, which, even with a strength

of one to one thousand, imparts a deep rose-red staining to sections within a few minutes. The staining is generally very diffuse, affecting the most diverse substances; even the red blood-corpuscles assume an intense rose-red shade, which is still deeper in sections made from alcoholic specimens, provided that the chromates have been used for hardening. The color is removed by absolute alcohol, at first very rapidly, then more slowly, so that, by the careful use of this agent, any desired shade can be produced.

A pure eosin-staining is very seldom of advantage (concerning the "eosinophil" cells, see below), although eosin is very often used for double staining, in connection with the coloring of nuclei. For this purpose the nuclei are, to bring out the contrast, best dyed blue, with gentian or methyl-violet (q. v.); or with hæmatoxylin.

A mixture of hæmatoxylin and eosin can be prepared, according to Renaut's suggestion, in which the double staining is effected at once. It is only necessary to add half a part of eosin to the solution of hæmatoxylin, already mentioned, in order to obtain such a mixture. The sections are usually too strongly stained with eosin at first; but, after a short stay in alcohol, the proper shade generally appears, whereupon it is better to examine the section in oil of cloves.

This method is, in a great many cases, the most convenient and the best for staining prepared sections. The nuclei of lymphoid cells generally appear to be most deeply colored, next those of the capillaries, of the other endothelia, and of connective tissue; while the nuclei of epithelial cells, etc., are less affected. Furthermore, differences appear in the depth of staining of the protoplasm, such as render possible the recognition of single elements, even with a low power. The parietal cells of the gastric glands, the giant cells of tubercle, etc., take a particularly deep color.

14. ANILINE-BLACK (NIGROSIN). ANILINE-BLUE. —These two dyes have a very similar action: they are used for staining axis-cylinders in sections of the nervous system. The section is immersed in a solution of about one per cent., in which it is deeply stained in a few minutes, and it is then washed in alcohol, and is again almost entirely decolorized. When the proper shade has been reached (which can easily be ascertained) the section is placed in oil of cloves or Canada balsam; the axis-cylinders and ganglion-cells take a very convenient blue or black tint, which is less marked in the neuroglia. They may be used also for staining protoplasm; quite weak solutions, for example, cause a very dark and characteristic coloring of the parietal cells (formerly called "peptic cells") of the gastric glands.

15. THE BASIC ANILINE DYES, WHICH STAIN NUCLEI.—The following basic aniline dyes, which stain nuclei, are the ones principally used:

Vesuvin (Bismarck-brown).
Fuchsin.
Gentian and methyl-violet.
Methyl-blue; also dahlia, magdala, methyl-green, etc.

These different substances possess nearly the

same peculiarities with regard to their action upon tissues, so that they may be considered together. They are all easily soluble in alcohol and in water, but their alcoholic solutions are only rarely used; we almost always employ the watery. It is better to keep on hand a concentrated watery solution which contains an excess of dye, and to filter off the necessary quantity from this fluid just before using. It is often convenient to have, instead of this quite opaque, deeply colored solution, a thinner one that acts less powerfully (about one to one hundred), to which, on account of its strength, a tenth part of its volume of alcohol may be added.

*Staining of Nuclei. Cells without Nuclei.*—If a section be placed in this solution it is stained an extremely dark color in a short time—within a few minutes. If it is washed in water and examined, we find a color that is almost completely diffused, is uniform, and is consequently useless. The advantages of the staining first appear after the action of the alcohol has taken place; the section when placed in a dish of alcohol discharges a large amount of dye, and is surrounded by a colored cloud; when re-examined after a few minutes, a beautiful, distinct staining of the nuclei is observed, and those of lymphoid cells, as well as those of connective-tissue and endothelial cells, have usually a much darker tinge than the nuclei of epithelial cells. The other substances are all nearly colorless, with some few exceptions to be mentioned shortly. The process is, therefore, soon at an end, the section being placed for a few minutes in the staining solution, and then for a few minutes in alcohol; from the latter it is either transferred to oil of cloves, in which the unstained portions become almost entirely cleared up, and is finally mounted in Canada balsam; or it is again placed in distilled water and is examined in glycerin. It should be remarked in this connection that the colors are well preserved in balsam-preparations, while in glycerin-specimens the nuclear staining is only retained when Bismarck-brown or vesuvin is used, while methyl-violet staining gradually disappears in glycerin (Weigert). For this reason Bismarck-brown is to be very highly recommended for most purposes, especially for the staining of fresh sections; these do not cause precipitates in solutions of Bismarck-brown, while they often occasion very troublesome granular deposits in a solution of methyl-violet.

By means of the distinct, separate staining of nuclei which can be produced by hæmatoxylin, picro-carmine, borax-carmine, lithium-carmine, etc., as well as by the aniline dyes already mentioned, Weigert was led to the important discovery that in a large number of pathological processes the cell-nuclei are— either invariably, or at least in certain types of cells— destroyed, a fact that had previously remained almost unknown. The absence of nuclei in the cells of the deep layers of the epidermis in small-pox was first discovered, then in diphtheria, later in various organs in the neighborhood of colonies of micrococci, in the looped tubules of the kidney in chromic-acid poisoning, in renal infarctions, in cheesy degeneration, etc. It was soon proved that the nuclei are absent in necrosed cells which have remained exposed for some time after death to a current (although diminished) of nutrient fluid; in many cases the protoplasm of the cells assumes simultaneously a glistening, homogeneous appearance, and these are the cases to which Cohnheim has applied the term "coagulation-necrosis." This expression has since been greatly misapplied; in my opinion it should be limited to those cases in which coagulation, as well as necrosis, has been demonstrated, or at least is probable; at all events, every cell in which a distinct nuclear staining does not occur should not be regarded as the seat of "coagulation-necrosis." It should be observed that when aniline dyes are used the nuclei of epithelial cells that are quite normal may, under certain circumstances, appear perfectly colorless if they are stained in an acid solution, or if dilute acetic acid as well as alcohol is used for decolorizing. Besides, it has not been positively proved that every cell in which a nucleus can not be demonstrated is therefore to be at once regarded as necrosed; it is always well in such doubtful cases of cells with non-staining nuclei, not to refer at once to coagulation-necrosis. Before the nucleus entirely disappears there are often found in its stead small, deeply-stained granules, which may be regarded as the products of disintegration of the nucleus, but which are often mistaken for micrococci by the unskilled; they are to be at once distinguished from these organisms by their very variable size.

The following bodies are stained by the above process, in addition to the nuclei:

1. The basement-substance of hyaline cartilage.

2. The mucous substances, as well as mucus in the glands.

3. Most of the micrococcus- and bacillus-forms; further mention will soon be made of these.

4. Certain protoplasmic granules, as the protoplasm of the so-called "food-cells."

FOOD-CELLS (Mastzellen). — The "food-cells," most thoroughly studied by Ehrlich, are somewhat bullet-shaped (sometimes flat and fusiform) bodies, which are about twice as large as lymphoid cells; they consist of a rather coarsely granular protoplasm, the granules of which are deeply stained with basic aniline dyes according to the method already described. The granules take a reddish color when the violet dyes are used; but the nucleus always remains unstained, and appears as a bright spot in the midst of the deeply stained protoplasmic particles. They are widely distributed throughout connective tissue, and are especially abundant in mucous membranes, in submucous and also in inter-muscular tissue, in serous membranes, etc., generally in the vicinity of vessels, but almost always singly. Their physiological and pathological significance is still very little known; it has not been proved that they bear any relation to nutrition. They are found in large numbers in slowly forming granulation, or connective-tissue growths, as in elephantiasis and in the neighborhood of tumors. They have also been found in leucæmic blood, but are absent from normal human blood.

THE STAINING OF AMYLOID WITH VIOLET ANILINE DYES.—The violet basic aniline dyes, methyl-violet and gentian-violet, show an interesting reaction with amyloid substance, as was discovered almost simultaneously by Heschl, Juergens, and Cornil; they stain this a deep red, while the nuclei are stained blue. The red color of the amyloid is preserved in dilute acids; in alcohol, on the contrary, it is at once discharged, while the similar reddish staining of the food-cells is retained in alcohol. We can not, therefore, decolorize specimens thus stained in alcohol in the usual manner, but we use for this purpose a dilute acid, such as acetic (one to one hundred); they must not be stained too deeply at the outset, hence a rather dilute solution (about one to one thousand) is used, which gives a sufficient color within a few minutes. The sections are examined and preserved in glycerin, in which, however, the nuclear staining is lost after a little while.

The difference in color between the red amyloid and the other blue portions is very striking; at the same time there is also in well-prepared specimens a distinct blue staining of the nuclei. In consequence, this process possesses great advantages over the dyeing of amyloid with iodine, or iodine and sulphuric acid, which was formerly the only one known. It is well known that by the systematic use of this method Eberth has learned the important fact that the cellular elements—for example, those of the liver, kidneys, and spleen—as well as the smooth muscular fibers, of the vessels, never undergo amyloid degeneration, but that this is always confined to the interstitial substance and homogeneous membranes. The aniline method is decidedly superior for the decision of such questions; on the contrary, the caution must be given that every red staining with methyl-violet does not necessarily denote actual amyloid degeneration; other hyaline bodies also, which are perhaps related to amyloid material, such as certain urinary casts, show the reaction, but they should not on that account be regarded as amyloid. The two kinds of amyloid which have been distinguished, by means of their different reactions with iodine and sulphuric acid, conduct themselves in a similar manner toward the aniline dyes.

Curschman has recently recommended methyl-green for amyloid staining; the amyloid portions become violet, while the normal tissue, especially the nuclei, are green. The difference is certainly very striking; however, the violet color of the amyloid only results from a chance mixture of methyl-violet with the impure commercial methyl-green. As a rule, the use of impure drugs is not to be recommended for our purposes; it is better to adhere to substances that are exactly defined chemically, and are in the purest condition.

IDENTIFICATION AND STAINING OF SCHIZOMYCETES. —The staining of schizomycetes has borne an essential part in the important discoveries which have been made during the last few years in the province of infectious diseases; there is, therefore, every reason why this process should be exactly described.

(*a*) *The Identification of Schizomycetes when unstained.*—In order to show schizomycetes when they are not stained, we generally take advantage of a peculiarity possessed by them, namely, their resistance to acids and alkalies. A section made from a fresh or alcoholic * specimen is rendered nearly transparent by strong acetic acid, or a dilute solution of potassium or sodium (about two per cent.). Among the few elements which resist this treatment the schizomycetes are at once recognized, by the characteristic form of the individual organisms. This applies particularly to the bacilli; in their case it is quite possible that minute crystals may give rise to error. In fact, the bacilli of typhoid fever and tuberculosis, for example, may thus be demonstrated very well in organs; it is better to employ, for this purpose, specimens which have been kept only a short time in alcohol, since, as a rule, the sections afterward cease to be perfectly transparent. Klebs and Baumgarten have laid particular stress upon this point; however, it is nearly always possible, even in the examination of old alcoholic preparations, to clear up the sections sufficiently, by warming them for a short time, under the cover-glass, with liquor potassæ or acetic acid, until bubbles begin to form.

Secondly, by the characteristic union or grouping of the individual organisms. This consists in the formation either of chains (diplococci, streptococci, strepto-bacteria, etc.), or of the so-called colonies (gliacoccus). It is possible only in very exceptional cases to mistake these for inorganic precipitates; they may be at once distinguished from very fine fat-drops by the

* Hardening in chromic acid, Müller's fluid, etc., is not suitable for the examination of micro-organisms, since by the action of chromium numbers of dark granules are formed in the tissues, which are very difficult to clear up.

fact that they resist the process of extraction with ether and chloroform.

It is easy to recognize the micrococcus-chains and groups by the peculiar glistening appearance of the separate granules, or by their color (brownish), which, moreover, is only observed in certain varieties, when several granules are placed one above the other, also by the nearly uniform character of the separate bodies, as well as by their clear, sharply-limited outlines under a high power. *We are perfectly certain that we have to do with organisms, if we succeed in proving the fact that they grow.* This it is possible to do if they develop in the interior of vessels, because in growing they distend the lumina unequally; since the rapidity of their growth varies·at different points, *varicose swellings* of the vessels result. This occurs very often in the blood-vessels (capillaries and small veins) in cases of metastatic pyæmia, ulcerative endocarditis, etc.; the author found a similar varicose injection of schizomycetes in the lymph-vessels in acute croupous pneumonia. These forms of "capillary emboli," resulting from unequal filling of the capillaries with granular material, were formerly observed, and it was then known that such emboli cause inflammatory foci; but this granular matter was described as "detritus" until Recklinghausen, and very soon after Klebs, Waldeyer, and others, led especially by the varicose form of the injection, made the important discovery that the "granular detritus" represents living parasitic organisms, or micrococci. For only a substance which attains a peculiar capacity for growth can cause such an irregular, nodular form of vascular injection. After this identification has once been assured, it is of course not necessary to show signs of growth in every single case in order to establish the diagnosis of a micrococcus-colony. *If we find, in a section made either from a fresh organ, or from one that has been hardened in alcohol, groups or chains of small granules, which are of nearly equal size, which resist treatment with alcohol and ether, as well as the energetic action of concentrated acetic acid and alkalies, even under the influence of heat, we are justified in calling these granules organisms.*

In all important, or at all doubtful, cases we shall of course prefer the staining-reaction, to be presently described; staining is naturally of great advantage also for finding schizomycetes in sections. I shall not discuss here the methods of cultivation, but refer to Koch's classical description in the communications of the Imperial Board of Health for 1881.

(*b*) *Staining of Micrococci, etc.*—Schizomycetes generally act toward dyes in a manner very similar to the substance of nuclei, from which, however, they are to be clearly distinguished by their resistance to alkalies, in which nuclei are at once dissolved. The methods of staining nuclei are therefore suitable for staining most schizomycetes also, and since we have here to deal with very minute bodies, we endeavor to obtain the deepest color possible; dyeing with hæmatoxylin and the nucleus-staining carmines is less adapted for this purpose than the basic aniline dyes (Weigert). We accordingly use very concentrated watery solutions of the latter substances, since the addition of a large quantity of alcohol injures their staining power; the intensity of the color may be still further increased, according to Koch, if the staining is carried on in the incubator at a temperature of about 50° C. The process requires from a few minutes to half an hour; the subsequent decolorization in alcohol (the alcohol must be pure, and above all perfectly free from acid) must not be too prolonged, since the color of the schizomycetes also is sometimes gradually removed by alcohol; the specimen is always mounted in oil of cloves, or in Canada balsam, and is examined by the aid of Abbé's illuminating apparatus, with the open condenser. It is generally of small importance which of the before-mentioned dyes is used; preparations stained with vesuvin or Bismarck-brown can be photographed, while the blue or red staining with gentian-violet or fuchsin usually presents the most striking and vivid pictures.

The stained micrococci (bacilli are unmistakable) might be considered either as the remains of nuclei, which, however, are generally to be distinguished at once by their variable size, and by their location at the site of the original nucleus, or as the granules of food-cells; but the latter are always easily recognized as such, by reason of their grouping around the central unstained nucleus. Many apparently insuperable obstacles to the recognition of these objects certainly do confront him who is not accustomed to accuracy in his work and observations. I have often noticed that an observer was inclined to regard as microcicci irregular, stained mucus-coagula, or other purely accidental objects. Others allow their sections to lie for days in impure water, instead of in alcohol. Numbers of schizomycetes then collect both on the surfaces, and at the edges of the section, so that, when the latter is subsequently stained, perplexities arise as to their meaning. There are also a number of minds which either never attain the capacity for the more delicate histological examinations, or succeed only after long and painful training.

Are all parasitic micro-organisms stained in this way? No; just as there is no one general method of histological examinations, so we can not expect to discover any single method of demonstrating micro-organisms. Micrococci and, so far as we know at present, all varieties of them, are stained in the manner described. Only the limitation must be imposed that they generally lose their power of absorbing or retaining coloring-matter after death. There are frequently found in the interior of organs, beside the deeply stained groups of organisms, much paler, even quite colorless, collections of granules, which may, with the greatest probability, be called dead micrococci. In the micrococcus-chains, also, differences in the intensity of the color are observed in the individual members of the chain, which are to be explained in the same way.

It is always possible that further study will reveal differences in the staining power of the various species of micrococci; as yet, however, nothing is known on this subject.

The micrococci of malignant endocarditis, of pyæmia, erysipelas, gonorrhœa, etc., act in quite a uniform manner with the methods which we employ.

It is well known that the micrococci of pneumonia are frequently distinguished by a peculiar capsule; this stains only slightly with gentian-violet and fuchsin, while the coccus itself assumes a very dark shade. The capsule and the coccus stain almost uniformly with Bismarck-brown and methyl-blue.

The spirilli of recurrent fever occupy a peculiar position. Most observers have not succeeded in identifying these *in situ* in the interior of organs. The spirilli are known to be differently constituted from most of the other schizomycetes, since they are quickly destroyed by acids and alkalies, and even by distilled water. They also resemble protoplasm in their behavior, rather than the substance of nuclei. They can not, therefore, as a rule, be colored by the ordinary process of nucleus-staining. Koch alone, the greatest master in this art, has succeeded in staining them, even in the interior of organs, by means of brown aniline dyes, and has photographed them. Even he declares that the identification of spirilli in hardened organs is a difficult task.

Bacilli act differently; the forms that usually appear during the decomposition of the body, as well as the bacilli of splenic fever, stain deeply. The staining power of typhus-bacilli was at first questioned, but wrongly; they only dye somewhat less intensely than the others, and even this slight difference completely disappears if the process is carried on in a warm chamber. On the other hand, small unstained spots, or apparent gaps, partly circular, partly elliptical, are found in the interior of the typhus-bacilli, which include about half the breadth of the bacillus, and sometimes more. These may possibly be the spore of the bacilli, or they may be "vacuoles." According to Goffky, another kind of sporeformation takes place in typhus-bacilli, when cultivated at the temperature of the body—namely, spores of the same breadth as the bacillus, attached together end for end.

The bacilli of leprosy stain with gentian-violet, methyl-violet, and fuchsin, but not with Bismarckbrown. In them, also, are found colorless spots similar to those in the bacilli of typhus fever.

(c) *Gram's Method.* — While, by the method which has been described (which until recently was the only known way, of staining schizomycetes), the parasites were colored in common with the nuclei, Dr. Gram, of Copenhagen, my esteemed friend and co-laborer, has lately succeeded in discovering a process of separate staining of schizomycetes, which may be regarded in most cases as the best, and even as the ideal method.*

In many instances, the deeply stained nuclei conceal the schizomycetes, which are always much smaller, and are, therefore, only slightly dyed. For this reason it is necessary, in the case of tissues that are rich in nuclei, such as the liver, spleen, lymph-glands, pneumonic lungs, granulation-tissue, etc., to make use of the old process of Weigert, and to employ very thin sections. Even then it is often quite difficult to see and to demonstrate the delicate objects. Gram's

* Gram, " Fortschritt d. Med.," 1884, S. 185.

method effects a *separate* staining of the schizomycetes. While everything else remains perfectly colorless, they, on the contrary, are dyed an intense blue, so that almost every individual in the section must at once strike the eye of the beholder.

The method is frequently purely empirical. Ehrlich's solution of gentian-violet in aniline water is used for staining. Some finely-powdered gentian-violet is shaken into a glass flask, to the depth of two finger-breadths; upon this pour aniline-water— that is, a saturated solution of aniline (a yellowish, oily fluid), which is at once formed by stirring four parts of aniline in one hundred parts of distilled water. The excess of aniline, which is partly precipitated and partly suspended in the water, so as to form a light cloud, is separated by filtration. The aniline-water, which is now perfectly clear, is poured upon the dry gentian-violet, and, in the course of twenty-four hours, after violent shaking, a saturated staining-solution is produced. We accomplish almost the same end by pouring about five parts of a saturated alcoholic solution of gentian-violet into one hundred parts of aniline-water. If the bottle is loosely stoppered with a funnel lined with filter-paper, the necessary quantity of the solution can be filtered off into a dish without further ceremony. The same filter can be frequently used, and the solution is preserved unchanged for several weeks, and even months. The section of the alcoholic specimen which is to be examined (or the cover-glass to which the dried preparation is adherent) is placed in this solution for a few minutes; the section must have previously been immersed in pure alcohol, not in water nor in dilute alcohol. The very deeply stained sections are then transferred to a dilute solution of iodide of potassium (iodine one part, iodide of potassium two parts, distilled water one hundred parts), when a dirty precipitate is formed. They are left in this for from one to three minutes, and are then removed to absolute alcohol. The color is soon completely discharged in the latter, and the alcohol is colored purple-red, while the sections become almost colorless. They are cleared in oil of cloves, and are mounted in xylol, Canada balsam, glycerin, or glycerin-cement.

While all of the tissue-elements appear perfectly colorless (sometimes a dull, bluish shade remains in the nuclei), the schizomycetes stand out as dark blue, almost bluish-black, granules or rods, threads, etc. It is at once evident that the micro-organisms are observed far more clearly, and in greater numbers, than by the former method of staining the nuclei. It is possible that we shall succeed in this way in discovering the probable parasites of those infectious diseases (as syphilis, typhus, dysentery, etc.) the pathogeny of which has hitherto been almost, or entirely unknown.

By the subsequent use of Bismarck-brown a very beautiful double staining may be produced, the schizomycetes remaining dark blue, while the nuclei become yellow or brown. In this way the relative positions of the parasites in the tissues appear more clearly, so that we learn whether they are situated within, or outside of, the cells.

It should be observed also that with this process the typhus-bacilli appear colorless, or are decolorized like nuclei, so that they are distinguished in this way from most of the bacillus forms. In many cases of pneumonia the cocci are almost entirely decolorized by the use of the iodine method; in most cases, however, they appear in enormous numbers, far more numerous than one would believe from using the old process of Weigert; a comparison of both methods applied to sections of the same series is very instructive. It must always be observed that there are cases in which the iodine method does not bring out the schizomycetes clearly, while they can be shown in another way.

(*d*) *The Staining of Tubercle-Bacilli.*—The peculiar relation of the tubercle-bacilli to aniline staining is of great importance. After efforts had long been made in vain to demonstrate anatomically the exciting cause of tuberculosis—the statements for and against the discovery of micrococci in tubercles were not well supported—Koch again succeeded, by the use of a modified staining process, in demonstrating the constant presence of a specific bacillus in the products of tuberculosis.* Since this concerns one of the most important discoveries of recent medicine, I desire to introduce here the original method of Koch, although it has been superseded by Ehrlich's modification (which was published soon after), and is practically only seldom employed. Koch's directions were as follows: The section (or the dry

---

* Koch, "Die Ætiologie der Tuberculose." "Berlin. klin. Wochenschr.," 1882, No. 15.

preparation) is immersed for twenty-four hours in a mixture consisting of two hundred parts of distilled water, one part of a concentrated alcoholic solution of methyl-blue, and one fifth of one part of liquor potassæ (ten-per-cent. solution). It is now stained dark blue, and is next placed for fifteen minutes in a concentrated watery solution of vesuvin. It is then washed in distilled water till the blue color has disappeared and a more or less dark brown shade remains; it is next dehydrated in alcohol, is cleared in oil of cloves, and is examined under a homogeneous lens, with the open condenser. The nuclei, as well as most of the varieties of micrococci, then appear of a brown color, as if they had only been exposed to the vesuvin solution; the tubercle-bacilli, on the contrary, are stained a deep blue. Koch was of the opinion that this depended upon the alkaline reaction of the staining solution, since the bacilli were never stained in acid or neutral dyes; the neutral solution of another dye then displaced the first staining everywhere except in the tubercle-bacilli, which retained the original color. There is also a specific reaction which affects the tubercle-bacilli only, in contrast with other schizomycetes; the bacilli of leprosy, according to Koch, behave in a similar manner, and they also retain the blue staining. The latter, however, are also dyed by employing the simple nuclear staining, with which the tubercle-bacilli stain only a little, if at all (see above). The modification which was soon after introduced by Ehrlich depends upon the following principles: In the first place, in order to render the staining solution alkaline, he does not use liquor potassæ, but another base, namely, aniline, a yellowish oily fluid, a saturated watery solution of which is able to dissolve far more of the dye than the dilute solution of potash. Then he employs strong mineral acids for decolorizing. He proceeds upon the principle that the tubercle-bacilli, which according to Koch remain colorless in neutral and acid dyes, but which are, on the contrary, stained by alkaline solutions of the same dyes, are surrounded by a cloud that is only permeable for fluids having an alkaline reaction. If, therefore, the staining has taken place in the alkaline solution, the nuclei, protoplasm, basement substance, etc., as well as the bacilli, are colored, so that the latter remain concealed. If, however, the specimen is placed in acid, the color will be removed from all the other elements, and in time even from the other micro-organisms that may be present in the preparation, since the acid bears a very close relationship to the dye. But the acid can not affect the bacilli, because their supposed capsules are impermeable to acids; the bacilli, therefore, remain as the only colored bodies in the midst of the otherwise perfectly colorless specimen, and thus stand out in the most striking manner.

Whether this hypothesis of the capsules of bacilli, impermeable to acid and neutral, but permeable to alkaline solutions, is right or not, the future must decide; at any rate, it explains most of the phenomena hitherto known, and has led to the discovery of a method through which the tubercle-bacilli, and, indeed, these only, are deeply stained in tissues and fluids.

The correctness of Ehrlich's hypothesis has been called in question by Ziehl; the latter also succeeded in staining tubercle-bacilli by adding phenol (a substance which generally acts like an acid) to aniline dyes. Furthermore, it was stated by Lichtheim,[*] Giacomi, and others that simple aqueous solutions of gentian-violet and fuchsin are capable, if sufficient time be allowed (the process occurs rapidly on the application of heat) of staining the tubercle-bacilli, though not so deeply as by Ehrlich's method. A large number of new methods of coloring bacilli have been published, but these are all, so far as they can be relied upon, only more or less important modifications of Ehrlich's. Since the latter, so far as is at present known, is perfectly reliable and has not been surpassed by any others, I shall confine myself to an accurate description of this process, which was at once accepted, even by Koch himself;[†] I myself have never had any reason to abandon it. An account of other so-called "methods," such as have been published, would be very uninteresting and without any practical value.

The following elementary principles should be remembered: While most of the other schizomycetes are rapidly and deeply stained by aqueous solutions of gentian-violet, fuchsin, Bismarck-brown, etc., this does not occur in the case of tubercle-bacilli; the continuous action of the dye is necessary, the specimen being kept in the warm chamber. Success in staining tubercle-bacilli deeply has not generally been at-

---

[*] "Fortschr. d. Med.," Bd. 1.
[†] "Mittheilungen aus dem k. Gesundheitsamt," Bd. 2, 1884.

tained by the use of Bismarck-brown; gentian-violet, or fuchsin, is used. Aniline-water—that is, a saturated solution of aniline in water—is employed as a solvent. When the staining has taken place, the specimen is treated with mineral acids, such as hydrochloric, nitric, etc., in watery or alcoholic solution. Those bacilli which retain their color after treatment with strong mineral acids are to be regarded as tubercle-bacilli. Their peculiar property of retaining their color in strong acids is perfectly characteristic of them; the diagnosis of tubercle-bacilli should only be regarded as confirmed if the staining has been retained after the use of acids. I hold that it is by no means allowable to omit the treatment with acids, simply on the ground of convenience. In properly prepared specimens all the other tissue-elements must be perfectly colorless, even the other varieties of schizomycetes which may be present, so that by completely extinguishing the structural image —that is, with the full Abbé's illumination—the bacilli remain quite alone in the field. In order to bring the tubercle-bacilli more easily into view, as well as to render clear their relative positions in the tissues, the nuclei may also be stained, in the manner described, with a color such as will offer the strongest possible contrast. The process is carried out in the following manner: The preparation of the staining fluid, gentian-violet in aniline-water, has already been accurately described (page 76). Fuchsin, or various other aniline colors (such as magdala, dahlia, methyl-violet, magenta, etc.), may be chosen instead of gentian-violet. It is immaterial whether the dye is added to the aniline-water in substance, or in a concentrated alcoholic solution. From ten to twenty drops, or more, of this solution, which can be kept on hand for weeks (although aniline-water itself decomposes pretty quickly when exposed to the light), are filtered off into a watch-glass; the section to be examined, or the cover-glass, is placed in the dish for about twenty-four hours. The deeply-stained section, or cover-glass, is transferred to a dish of distilled water, in which it is washed; then it is again placed in a watch-glass containing a solution of three parts of nitric acid and one hundred parts of alcohol. In a very short time—from three to five minutes at the utmost—the decolorization has reached the proper degree, the specimen is removed to pure alcohol, and is then examined in oil of cloves; the cover-glass preparations may also be washed and examined at once in water. If a double staining is desired, Bismarck-brown is employed for specimens that have been colored blue, methyl-blue for those that are red. Care should be taken that the nuclei be not too deeply stained, lest the tubercle-bacilli should be concealed.

The decolorization can also be effected by means of a watery solution of nitric acid, but this must be considerably stronger than the other (about one to three), in order to act in the same time; the more dilute alcoholic solution is to be preferred, since it does not attack metallic instruments. If the decolorization is not complete, a portion of the former dye again appears after washing off the acid; this remnant is then easily removed by the combined action of the acid and alcohol. The staining of the tubercle-bacilli, on the contrary, persists even after it has been exposed for more than a quarter of an hour to the action of the acid solution.

We are indebted to Rindfleisch for an essential improvement in the method—namely, rapid staining in the presence of heat. The watch-glass filled with the staining solution, which contains the specimens, is placed in a thermostat which is kept at a temperature of 60° to 80° C., or better, it is held over the flame of a spirit-lamp or a small gas-jet, and heated until thick vapors arise; then the staining of the tubercle-bacilli occurs in a few minutes. This process has rapidly come into general use, especially in the case of cover-glass preparations (sputa); it is less suitable for sections. The further manipulations, decolorizing, or double staining, are then effected in the usual manner, at the temperature of the room. The bacilli of tuberculosis may be seen in this way with a power of three hundred, and often even without the use of immersion-lenses. The best objectives are requisite in every case, however, in order, in the more difficult cases, to see the fine bacilli which may perhaps be concealed. The careful observer will only give a positive opinion concerning the absence of bacilli in a given specimen (which opinion may often be extremely important) after he has exhausted the best optical aids at present known; otherwise, gross errors would occur, as has already happened in the experience of many. He who desires to undertake the examination of tubercle-bacilli for diagnostic purposes must accordingly be provided with an Abbé's illuminating apparatus, and a powerful immersion-lens—if possible, an oil-immersion.

Unfortunately, the preparations are generally not permanent; the color of the bacilli is apt to fade or to disappear entirely within a few months, often, still more rapidly.

The tubercle-bacilli may be identified even without staining. Baumgarten demonstrated them in the inoculated tuberculosis of animals, and later in man also, by treating the sections with liquor potassæ, even before he was aware of Koch's results. In fact, it is possible in many cases of tuberculosis to identify the bacilli in this simple manner; however, we must necessarily prefer, in every case, Koch's staining process, with Ehrlich's modification. For by the aid of the latter alone do we decide whether we really have to do with tubercle-bacilli, since, as already mentioned, the other varieties of bacilli, with the single exception of those of leprosy, do not manifest this special behavior toward the staining process. Besides, it is an undoubted fact that in many cases in which the bacilli are few in number, their discovery is rendered much easier, or possible, by their intense staining. In the tubercles of fungous joint-disease, for example, we shall seldom succeed in seeing the individual bacilli by the use of the potash-method, whereas we almost always succeed with the staining. Moreover, we can never dispense with the staining-process in examinations undertaken for diagnostic purposes (sputa, pus, etc.). The colored tubercle-bacilli frequently show interruptions, in the form of clear, unstained portions of round or oval shape, which are regarded as spores.

16. THE NOBLE METALS. (*a*) *Silver.*—The "silver method," introduced by von Recklinghausen, is of great value in normal histology; the important discovery that the walls of lymph- and blood-vessels, formerly regarded as homogeneous, are really composed of endothelial cells, was rendered possible through the use of the silver process, and it is still almost indispensable for the demonstration of this fact. It is but seldom used for our purposes; the most simple case is that in which it is a question of demonstrating an endothelial layer upon a given surface, but it is more difficult if pathological changes of the walls of blood- or lymph-capillaries are to be studied.

The difficulties of the method consist in the fact that the silver salt which is employed—usually the nitrate—forms, with the albuminous fluids of the body, granular and thready coagula, which are of very irregular shape, and may readily form deceptive images; in order to avoid this we endeavor to confine the silver precipitate entirely to the borders of the endothelial cells. The silver process can therefore be used only with normal surfaces; the agent invades to a very slight extent the interior of the tissues. It is better to use quite dilute solutions, as one to five hundred; the surface is, if necessary, washed with distilled water or with a dilute solution of nitrate of sodium (two per cent.), then the solution of silver is poured upon it, and after about a minute it is again washed with distilled water. After a short interval (most rapidly under the influence of sunlight) deep black lines appear at the borders of the endothelial cells; the nuclei are not usually colored, but they can afterward be stained with hæmatoxylin. The precipitate of silver is readily soluble in dilute ammonia. In order to define the borders of the endothelial cells in the capillaries the silver salt is injected into an artery; by injecting the solution of silver into the bronchial-tree, the borders of the alveolar epithelia of the lungs are stained. If about five per cent. of gelatin is added to the solution while heating, we obtain a "silver cement," also very useful for injections, which stains the borders of cells lining injected cavities a brown color. If a cornea is placed for a short time in a silver-solution, or if it is even rubbed with the solid stick, there occurs a dark brownish staining of the basement-substance, within which the corneal corpuscles appear as bright radiating figures, or spaces.

It should always be observed that the silver process is only applicable to fresh preparations, in which cadaveric decomposition has not yet begun.

(*b*) *Gold.*—The chloride-of-gold method, which was introduced by Cohnheim, offers similar difficulties; this reagent, too, penetrates only to a slight extent into the interior, and has besides this disadvantage, that its action is not quite constant. The conditions under which the reduction of the salt, and at the same time the staining, take place are not yet exactly known to us. However, successful gold preparations are very valuable on account of their extremely exact delineation, so that we can not dispense with this process in experiments on the cornea, keratitis, reproduction, etc.

The advantages of the gold method consist in this: 1. The protoplasm of cells, especially in the cornea, is very darkly stained, and is thus sharply distinguished from the perfectly clear basement-substance. 2. The axis-cylinders of the nerve-fibers are stainèd separately. Different solutions of chloride of gold are used, from one to one hundred to one to one thousand; the cornea, or other lamellæ that are to be stained, are left in the solution from ten minutes to an hour; this reagent also affects only the superficial portions of the specimen. They take a straw-yellow stain in this, and are then placed for a somewhat longer time (about twenty-four hours) in a dilute acid, as acetic, formic, tartaric, or citric; Ranvier has recently employed lemon-juice. Then the reduction has either occurred, already or it takes place completely in the course of several days, during which time the specimen is preserved in alcohol or glycerin. The color is dark violet. The preparation acquires a firm consistence through the action of the gold salt, so that it may be cut into the thinnest sections; or it may be still further hardened in alcohol. Chloride of gold may also be used for the staining of sections of the nervous system, which have been prepared after the ordinary treatment with Müller's fluid (Leber); the sections are immersed for about an hour in a one-half-per-cent. solution, and are then placed in distilled water, when they show in the course of one or two days a deep violet staining of those parts which contain the normal nerve-medulla.

The method is, therefore, very useful for identifying degenerations and atrophies in the peripheral nerves, and in the white substance of the central organs.

In order to stain the nerves with gold in alcoholic preparations, Frisch recommends the following method: The sections are washed in water, are placed for twenty-four hours in a solution of sodium chloride (six-tenths of one per cent.), then for ten minutes in formic acid (ten per cent.); they are next thoroughly washed for from one-half to three hours in a one-per-cent. solution of sodium chloride and gold, while protected from the influence of light. Then they are again washed and immersed in ten-per-cent. formic acid for twenty-four hours.

(c) *Osmic Acid.*—This reagent, which was first used by Max Schulze, has been employed in many different ways. It serves (1) for fixing and hardening delicate tissue-elements in nearly their natural form; (2) for bringing out or staining fats, including the nerve-medulla.

The osmic-acid solution also penetrates only the superficial layers of preparations. A solution having a strength of from one-tenth to one per cent. is used. It should be remembered that the fumes of osmium are very irritating to the conjunctiva and mucous membrane of the nose. The solution of osmic acid should be kept in brown bottles, as well as that of the chloride of gold and nitrate of silver.

In small bits of fresh tissue, which have been immersed for a short time (about an hour) in dilute osmium solutions, and have then been mounted in glycerin, the cellular and fibrous elements are often isolated very well, since they have received through the action of the reagent a certain resistance as well as a brown color. This method is to be especially recommended in the case of the nervous system; the nerve-medulla is stained dark blue or black. The red blood-corpuscles are also colored brown by osmium, and then become quite resistant toward most influences; even the fumes of osmium, which are given off at the temperature of the room, act in this way. The latter reaction is effected by holding the specimen, which is on the under surface of the slide, over the neck of a bottle filled with osmic acid. If the acid acts more vigorously and for a longer time, small pieces of tissues, as nerves, etc., become hardened. The various fats, as well as the medulla of nerves, are stained a deep-blue color in a few minutes by the action of the acid, through the reduction of the metal, as is generally stated; a special combination probably occurs here. This striking dye is of great value to us, since it gives the best staining-reaction for fats. The reaction takes place very well, even in sections that have been made from alcoholic specimens; that part of the fat which is not dissolved by alcohol is stained dark brown within a quarter of an hour. The osmium process can be highly recommended for the preparation of permanent specimens of fatty degeneration of the kidneys, liver, heart, granulation-tissue, and tumors, for purposes of demonstration. The sections should be mounted in glycerin.

17. AMMONIUM SULPHIDE. (*Siderosis.*)—An aqueous solution of sulphide of ammonium has been used

extensively by Quincke in pathological and histological examinations as a test for iron;* the iron, contained in the interior of cells in the form of an albuminate, is precipitated by the ammonium sulphide in the form of dark-green granules (sulphate of iron). The ferruginous particles are often recognized by their yellowish-brown color before the action of the reagent occurs, so that this is not indispensable, and, on the other hand, all yellow pigment-granules do not show the dark green staining with the reagent. The substance formerly employed as a micro-chemical test for iron, ferrocyanide of potassium mixed with hydrochloric acid, is less favorable because it coagulates albuminous bodies, and, besides, gives colors that are readily diffusible; the Berlin blue which is formed is not quite insoluble in the acid albuminous fluid. The dark green staining of the ferruginous granules with ammonium sulphide appears in a few minutes in sections of alcoholic specimens, and lasts for weeks.

Normal red blood-corpuscles do not give the reaction, hence the inference is that iron is not precipitated from all of its combinations by means of sulphide of ammonium. On the contrary, Quincke states that, even in the normal liver, and especially in the spleen and medulla of bones, "siderosis" is present—that is, ferruginous particles, which must have been derived from disintegrated red blood-corpuscles, can be identified by the aid of the sulphide. The destruction of red blood-corpuscles increases

* Quincke, "Über Siderosis," "Deutsch. Arch. für klin. Med.," Bd. 25.

very greatly as a result of transfusion (in so-called artificial plethora), so that the physiological siderosis becomes considerably extended. The particles of iron in the liver are contained within the capillaries in the interior of white blood - corpuscles; in the spleen and in the osseous medulla they lie within the cells of the pulp. A very decided siderosis also occurs under similar conditions in man—that is, in cases in which a marked destruction of the red blood-corpuscles occurs, especially in pernicious anæmia. Iron is then demonstrable in the cells, capillaries, and also in the perivascular connective-tissue of the liver, in the gland-cells of the pancreas, in the epithelium of isolated looped tubules of the kidney, and also in the spleen and osseous medulla.

## IV.

## OTHER METHODS OF PREPARATION.

PRESERVATION OF SPECIMENS.—Most of the methods of preparing specimens have already been mentioned in the former section; the processes of hardening in alcohol and chromates, and of decalcification with mineral acids, have been discussed under these reagents. It remains to present a few of the special methods.

1. BOILING.—It was formerly the custom to boil anatomical specimens occasionally, in order to prepare them for histological examination; but the method was first employed in a rational way by Posner [*] (from a suggestion made by the late Perls), especially with the view of precipitating the dissolved albumin rapidly and surely *in loco*, and thereby rendering it visible.

The pieces of organs, about the size of a hazel-nut or walnut, are thrown into boiling water, from which they are removed in a few minutes, and washed in cold water. They have then, as a rule, a somewhat tough, elastic consistence, and can be at once cut with the razor, or they may be perfectly hardened in alcohol. The coagulated albumin appears in such prepa-

[*] Posner, " Virch. Arch.," Bd. 79, S. 311.

rations, as a coarsely granular mass; most of the cell outlines also have become sharp and distinct.

The method possesses peculiar advantages for the examination of kidneys in cases of albuminuria and of oedematous lungs. If such organs have been hardened simply in alcohol, it is also possible to identify the granular coagulated albumin in the interior of the Malpighian capsules, or alveoli, especially in the superficial portions of the specimen, which were most directly exposed to the action of the alcohol. However, this is effected much more perfectly by means of the boiling process, by which a rapid and complete coagulation is attained in a very short time. Aside from this action, most structures are only slightly changed by a brief exposure to a boiling temperature.

The boiled specimens can also be cut with the freezing microtome.

2. THE DRYING of preparations, in order to render them suitable for cutting, was once much practiced. Since the general introduction of the art of hardening in alcohol (in the case of porous or very soft objects, after previous saturation with mucilage), drying has been almost entirely given up. Preparations shrink a great deal during the latter process; but the sections swell up again in water, although very irregularly.

For the purpose of macerating or isolating certain tissue elements, we employ

3. ARTIFICIAL DIGESTION, which is occasionally used for pathological examinations, but more especially in normal histology. We use pepsin or tryp-

sin—that is, artificial gastric juice, or extract of pancreas.

Artificial gastric juice is best prepared from the mucous membrane of the fundus of the pig's stomach. The pieces of mucous membrane are chopped fine, and are soaked for about an hour in very dilute hydrochloric acid (one to one thousand), which is placed in an incubator kept at the temperature of the body; the mixture is then filtered. Commercial pepsin may also be used; but, in any case, the digestive power of the fluid must be tested on bits of fibrin, or a piece of loose, coagulated albumin. These should be dissolved in a short time.

The extract of pancreas is prepared as follows:[*] The pancreas of a freshly killed bullock is chopped in pieces and completely extracted with alcohol and ether in an extracting-apparatus; the white, friable mass that remains after the evaporation of the ether is digested for about four hours, at a temperature of 40° C., with five to ten times its weight of salicylic acid (one-tenth of one per cent.), or with distilled water, and is then strained and filtered.

The artificial gastric juice digests in a short time (at the temperature of the body) connective tissue, muscle, most cellular elements, etc., while elastic tissue and nerve-fibers resist it. Extract of pancreas, on the contrary, or the trypsin which it contains, dissolves in an acid fluid elastic fibers, as well as the fine fibers of the neuroglia, while the connective-tissue fibrillæ remain intact.

[*] Kuehne, "Verhandl. des med.-naturf. Vereins zu Heildelberg," I, 1877.

Digestion may be carried on either in the incubator, or (by means of the warm stage) in the microscopic specimen itself, under the eyes and the constant control of the observer. It has been recently discovered, by the aid of this method, that the gray fibrillary substance, which is found in such quantities in the posterior columns of the spinal cord in tabes, corresponds perfectly in its chemical composition with the fibers of the neuroglia; it is also rapidly dissolved, while the fibers of the pia and their processes remain intact (Waldstein and Weber).* The so-called neurokeratin of Kuehne resists trypsin perfectly; it remains in the interior of nerve-fibers in the form of a delicate net-work, when these are extracted with hot ether and chloroform, and then with trypsin. The horny net-work was regarded by Kuehne, and by many of his followers, as a preformed structure—"the horny basis of nerve-fibers." Hesse and others oppose this view, more recently also Waldstein and Weber (pupils of Ranvier).

These authors assert that neurokeratin is mingled in quite a diffuse manner with the medulla, and that it only assumes its peculiar net-like structure during the process of extraction; the form of the net-work can be varied at will according to the variation of the process. By similar treatment the same "horny network" can be produced in the irregular drops of the escaped medulla (myelin) as in the interior of nerve-fibers. Neurokeratin disappears in degenerat-

* "Arch. de physiolog. norm. et pathol.," II. Reihe, Bd. 10, 1882, S. 1.

ed nerves, as well as in gray degeneration of the white substance of the central nervous system.

4. IMBEDDING.—Most of our hardened specimens are cut without imbedding. In speaking of the microtome, allusion was made to the little devices of gluing the preparations upon cork, and of securing them between pieces of hardened liver; we are thus able in most cases to fix the specimens sufficiently without further imbedding, so as to obtain perfect, even sections. If we have an irregular surface to cut, and it is desirable that the sections should include the surface itself, as in the examination of a menstrual mucous membrane, a thin layer of mucilage is applied to the surface in question, and upon this is placed a slice of hardened liver. The gum soon hardens in alcohol, so that a firm union between the liver and surface of the preparation occurs. If the inequalities of the surface are considerable, the mucilage is less useful, since in thicker layers the gum acquires a stony hardness, so as to injure the knife; glycerin-cement is used in this case, after the example of Klebs. This is prepared in the following manner: Ten grammes of the finest well-washed gelatin are allowed to soak in distilled water, the residue of water is poured off, and the swollen glue is dissolved by gentle heat; to this add ten grammes of glycerin, and also a few drops of phenol to prevent the formation of mold. A small part of this mass, which becomes solid at the temperature of the room, may readily be dissolved by warming at each time of using; the irregular surface of the preparation is covered with the fluid,

and assumes a proper consistence for cutting when placed in alcohol.

In other cases, however, we are compelled to cut a specimen *in toto*—that is, to include all the surfaces of the same in the section—a task which very often devolves upon zoölogists and embryologists, but upon us only in exceptional cases. The specimen must then be surrounded with a solid mass; glycerin-cement can also be used for this purpose, but it is better to employ a substance which does not shrink when it hardens, and even when it is treated with alcohol. We use, for example, a mixture of equal parts of wax and oil, or two parts of spermaceti and one part of oil of bergamot, or the following:

    Paraffin . . . . . . 5 parts.
    Spermaceti . . . . . 2 "
    Suet . . . . . . 1 part.

The mixture liquifies when heated gently, is poured around the alcoholic preparation (which has been placed in a tin box, properly adapted to the microtome-clamp), and, after cooling and an additional stay in alcohol, it again solidifies. If we wish the fat to penetrate the specimen, the latter must of course be completely dehydrated; it is usually immersed in an ethereal oil, as oil of bergamot, before it is saturated with the fat. In this way the specimens acquire an exceedingly even consistence; the fat is afterward removed from the sections. As is apparent, the process is somewhat lengthy; it will be very seldom necessary to employ it for our purposes.

Another favorite imbedding material is Caberla's.

This consists of fifteen parts of white of egg and one part of a ten-per-cent. solution of carbonate of sodium; the accompanying yolks are added, and the mixture is shaken. The object is placed in a paper box, which stands in a dish filled with alcohol having a strength of eighty per cent., and the dish is then heated on the water-bath up to about 75° C.; after warming for half an hour, coagulation has occurred to a sufficient extent, and the specimen is hardened in alcohol.

These imbedding substances are opaque, so that marks must be placed on their surfaces in order to show the position of the preparation.

The most valuable imbedding material is certainly celloidin, for the introduction of which into microscopical technology we are indebted to Schieferdecker ("Arch. für Anat.," 1882). Celloïdin is a substance resembling collodion, and is sold in solid masses. It dissolves slowly in a solution of alcohol and ether, forming a sirupy liquid. If the alcoholic preparation is now immersed in this, it becomes thoroughly saturated with the celloidin solution in the course of several hours (about twenty-four). The specimen thus filled with celloidin is then placed in seventy or eighty per cent. alcohol, in which the celloidin again solidifies, so that the specimen has a very uniform consistence for cutting. It is next transferred to the microtome, enveloped as it is in a nearly transparent mantle of celloidin, so that every section is surrounded by a like covering, and can be stained, examined, and mounted together with this. Oil of cloves is not used as a clearing agent in this case, since it dissolves celloidin, but oil of origanum or cedar.

The celloidin method has been approved in many instances; it is especially valuable for eyes, the spinal cord, etc.

5. PROCESS OF INJECTION.—This is employed far less frequently in pathological examinations than in normal histology.

It is highly important that we should study the *natural injection* of the blood-vessels with blood, as well as the lymph-vessels with lymph, and we seldom resort to an artificial filling of the lumina. Hence, only a brief sketch of the rather complicated process of injection is presented, while those who are interested in the subject are referred to the well-known text-books of Ranvier and Frey.

(*a*) *Injecting Material.*—We use as an injecting material a transparent but deeply colored fluid, such as solidifies within the vessels, generally a glue; in using the latter, the injecting fluid and the organ to be injected must be raised to rather a high temperature (40° or 50° C.). For this purpose large tin dishes are used, which are filled with water and heated from beneath. Glue injections are therefore rather troublesome, but they are preferable to watery solutions, on account of the stability of the material.

Soluble Berlin-blue, which can be obtained from druggists, serves for coloring the mass; sometimes this only dissolves after the addition of a little oxalic acid. A solution of this material in from ten to twenty parts of water may be used directly for injecting, or five parts of alcohol and five of glycerin may be added. Or, on the other hand, the hot, watery solution of Berlin-blue may be poured gradually into the same quantity of hot, concentrated solution of glue, while stirring constantly; the latter is prepared by allowing fine, well-washed sheets of glycerin to soak for one or two hours, at the temperature of the room, in about double the quantity of distilled water; the swollen glue is then liquefied by gentle heating upon the water-bath.

Since the "soluble Berlin-blue" of the druggists is not always reliable, we shall repeat here the exact direction of Thiersch for preparing the substance independently.*

*Thiersch's Berlin-blue, prepared with Oxalic Acid.*—Make a cold, saturated solution of sulphate of iron (A), a similar one of ferrocyanide of potassium (B), and, thirdly, a saturated solution of oxalic acid (C). Finally, a warm, concentrated solution (two to one) of rather fine glue is needed. About fifteen grammes of the solution of glue are mixed with six cubic centimetres of solution A in a porcelain dish. In a second larger dish thirty grammes of the solution of glue are mixed with twelve cubic centimetres of solution B, and to this twelve cubic centimetres of the oxalic-acid solution C are subsequently added. When the mass in both dishes has cooled to about 25° or 32° C., the contents of the first dish are added, drop by drop, to the mixture in the second while agitating constantly. After complete precipitation the resulting dark-blue substance is heated for some time at a temperature of 70° to 100° C., while

* From Frey, " Das Microscop," 1881.

stirring, and is finally filtered through flannel. The injections of Berlin-blue are beautifully colored, but, in the course of time the color gradually disappears as a result of reduction; the blue dye is again restored through the action of an ozone-carrier, as oil of turpentine.

Injections of carmine, on the contrary, are quite permanent; but here we encounter the difficulty that, in using an alkaline solution of ammonia-carmine, the red color at once transudes; the solution must, therefore, be neutralized. The neutralization must be effected with extreme care, since the material otherwise becomes opaque and perfectly useless, by reason of the coarse precipitate of carmine.

*Cold Fluid-injection of Carmine.*—One gramme of carmine is dissolved in a little water with one gramme of ammonia, and to this mixture twenty cubic centimetres of glycerin are added. To this solution is added carefully a mixture consisting of twenty cubic centimetres of glycerin and one cubic centimetre of hydrochloric acid, while stirring violently; the whole is then diluted with forty grammes of water (Kollmann).

*Frey's Carmine-cement.*—Take a solution of ammonia and one of acetic acid, the number of drops of which necessary for ready neutralization have been previously determined. From two to two and a half grammes of the finest carmine are dissolved in a dish by stirring with a fixed number of drops of the ammonia solution (which can be increased or diminished at pleasure), and with about fifteen cubic centimetres of distilled water, and the solution is then filtered;

several hours are required for this purpose and a loss of ammonia results through evaporation. The alkaline ammonia-carmine solution is added to a filtered, moderately warmed, concentrated solution of fine glue, while agitating; the mixture is heated for a while upon the water-bath, the number of drops of acetic acid necessary to neutralize the original solution of ammonia are added slowly, and the whole is stirred. Thus the carmine is precipitated in the acid solution of glue.

In injecting organs that have a pretty strong alkaline reaction, a small quantity of acetic acid may still be added to the material. During the injection the temperature should not exceed 45° C.

Other substances also have recently been employed for injecting the finest lymph-spaces, such as oily fluids which are colored with alkanet, generally oil of turpentine, or even chloroform in which a dark resinous body, as asphalt, is dissolved.

(b) *Injecting Apparatus.* — An injecting-syringe is often employed, which must work well and be very carefully cleansed after using; the syringe is connected with the canula, either directly or by means of a rubber tube. The syringe and canulæ are made of metal or glass; the latter we can easily prepare for ourselves of any desired shape. The piston of the syringe must not fit too tightly, and must move quite smoothly, without jerking.

Injection under constant pressure is an undertaking rather more complicated, but is one to be highly recommended. If the large machine of Hering is not at hand for this purpose, the necessary apparatus can be easily arranged by means of some wash-bottles and rubber tubes. The bottles are closed by rubber stoppers, perforated in two places, each of which is armed with two glass tubes in the ordinary manner, the short, thick one ending just below the stopper, while the other long one reaches down to the bottom of the bottle. One bottle, $A$, is nearly filled with the injecting-fluid; the long glass tube which dips into the fluid is connected at its other end with the injecting-canula. The other short tube is connected with the second bottle, $B$, which serves as an air-chamber, and at first contains nothing but air. This second bottle is again attached, by means of a long rubber tube and the glass tube that extends to the bottom, with a pressure-vessel, $C$, which can be elevated at will (by placing blocks under it) and is filled with mercury. If, now, the mercury is allowed to flow from the pressure-vessel, $C$, into the bottle, $B$, which represents the air-chamber, the air-space in $B$ is subjected to a pressure corresponding to the difference in level of the mercury in the two vessels; this pressure is at once transmitted to the injecting-fluid contained within the bottle $A$, and it is forced into the lumen of the vessel under the same pressure. Care is to be taken that the difference in level of the two portions of mercury remains the same. If the pressure-bottle, $C$, is elevated according as the fluid gradually flows out, the injecting-pressure remains constant. If there is a bulb-apparatus at hand, this may be substituted for the pressure-bottle. The pressure-vessel can be filled with water instead of mercury; in order to attain a greater pressure, the latter is placed upon a cupboard, etc. Of course, all the connections must fit accurately and in an air-tight manner. The canula, which is tied in the vessel, is filled with the injecting-material, or with distilled water, in order to avoid impurities, and is attached to the syringe, or other apparatus, in such a way that no air-bubble shall enter; the injection can then be begun. Moreover, the stream of injecting-fluid must never be interrupted by air-bubbles; they can certainly be avoided by carrying out the preparations carefully.

If the injection is taking place, the fluid soon escapes from the lateral or adjacent branches, especially if the organs, or portions of organs, have been removed at an autopsy before the injection was contemplated; the vessels must then be tied. It is better to leave the veins open at first, so that the blood can escape; these may also be ligated before the process is finished. By choosing proper arterial branches valuable injections can be made even in kidneys, lungs, livers, etc., which have already been cut through; this will be limited, however, to portions of the organ.* The injection is discontinued when the staining of the organ is sufficiently intense; the increase in the consistence of the parts also furnishes an indication as to when it is necessary to stop. After being injected, the specimens are at once transferred to alcohol.

During the injection of organs that have been

* Rindfleisch uses for this purpose thin elastic catheters, which are inserted deeply into the interior of organs, like injecting-canulæ; small openings are made at their extremities.

subject to pathological changes, we must frequently contend with extravasations; these can be disregarded, at least in many cases, by avoiding excessive pressure. In injecting lymph-vessels we frequently use delicate needle-canulæ (the so-called "puncture-injections"), which are carefully thrust into the organ at the desired spot; an "extravasation" necessarily occurs in the neighborhood of the injection, but the lymph-vessels are often filled most beautifully, and that, too, in a very simple and rapid way.

The methods of performing injections for physiological purposes have been greatly improved and extended during the past few years by Cohnheim, Heidenhain, Arnold, Thoma, and others. These have attained much importance in connection with many questions in pathology, especially in the case of the kidneys. However, they are only important in experimental investigations; we can not enter into a discussion of them here.

6. PRESERVATION OF SPECIMENS.—In order to preserve for a few days fresh specimens that lie in salt-solution, it is only necessary to place them in a moist space. This is very easily arranged in the following manner: A large, flat dish is filled to the height of several millimetres with water; in the dish stands upon three feet (formed of corks attached to it by means of sealing-wax) a small glass plate, which supports the specimens; a bell-glass, lined on its interior with damp blotting-paper, is inverted over this so that in this way the space in which the specimens are preserved is kept closed and sufficiently moist. Instead of the glass plate a wooden or metal *étagère*

may be used, in which the preparations are placed in several layers one above the other without touching; or a number of these plates or tables may be built up, one over the other, an antiseptic being added to the surrounding water. Fresh specimens do not preserve their perfect delicacy very long; it is of no use to seal them up in salt-solution, since the elements are soon destroyed, as a rule; and besides, the water generally evaporates in spite of care in cementing. Many objects can be preserved in acetate of potash, a saturated solution of which is known to be stable in the air; but in this, too, much of the original delicacy of the outlines is lost.

The most important question is the preservation of sections made from alcoholic specimens. If the sections are mounted in glycerin, they are already sufficiently preserved, and it is only necessary to fasten the cover-glass so that dust, or any other impurity, which collects upon it, can be wiped away. For this purpose the superfluous glycerin is pressed out by placing a bullet upon the cover-glass, and is absorbed by blotting-paper; then the slide is carefully cleansed around the cover-glass—that is, every adherent trace of glycerin is removed, which is best done by means of fine linen moistened with alcohol. The edge of the cover-glass is then surrounded with some hardening cement, and is thus fixed to the slide. Canada balsam in chloroform, or Brunswick-black, mask-varnish (Maskenlack), etc., is used as a cement. Following Ranvier's suggestion, I have long employed a thick solution of good red sealing-wax in alcohol. For glycerin we may substitute glycerin-cement, which liquefies on warming slightly and hardens on cooling. Or gum-arabic is added to the glycerin, so that the layer at the edge gradually becomes solid. Farrant's mixture, consisting of equal parts of glycerin, gum-arabic and a saturated solution of arsenious acid, is worthy of recommendation. For most cases simple mucilage is sufficient; the specimens are inclosed in the sirupy liquid, the latter dries up at the edge, and without any further care the sections are permanently fixed and preserved.

Preparations that have been cleared in oil of cloves are mounted in Canada balsam, as was previously stated (page 40); here, too, further cementing is unnecessary.

## V.

## THE OBSERVATION OF LIVING TISSUES.

THE CIRCULATION. INFLAMMATION.—The observation of pathological processes in living tissue by means of the microscope is practicable only to a very limited extent in man; the instrument devised by Hüter for observing the circulation in the mucous membrane of the cheek, and in similar regions, furnishes very defective images. In warm-blooded animals, also, the difficulties are considerable; however, these have been overcome, for Stricker and Thoma have constructed complicated instruments which allow of the observation in mammalians of the circulation, of its pathological disturbances, and of inflammation. Systematic artificial respiration is necessary, otherwise the curarized animals would soon die; beside this, the exposed transparent part which is being observed microscopically (the mesentery is the best) must always be kept at the temperature of the animal's body. All these precautions, however, have not hitherto led to any very important results; our knowledge has not been essentially advanced through careful investigations made upon warm-blooded animals. The well-known, striking processes were first demonstrated in the cold-blooded animal (the frog), principally by Cohnheim. On account of the great importance of the observations in question the very simple methods necessary for making them will be briefly described. In order to observe the circulation and its disturbances in frogs it is well to paralyze the voluntary movements of the animals by means of curara. If a particle of curara one half to one millimetre in diameter is placed under the skin of a large frog, the animal becomes motionless in the course of half an hour, while the vegetative functions continue. The frog's small need of oxygen can be supplied for days through the cutaneous respiration alone. Three regions in particular can then be utilized for the study of the circulation.

1. *The Web.*—The web possesses the advantage that it is not necessary to inflict an injury in order to observe the vital processes, since it is enough to simply separate two toes and to fix them apart. This part is therefore very valuable for many observations; but it is inferior in transparency to other objects soon to be mentioned. Although animals are selected which are as poor in pigment as possible, the pigment-cells, as well as the sharp outlines of the numerous layers of pavement epithelial cells, are troublesome. Besides, real inflammatory processes extend merely to a slight distance in the dense tissue; only disturbances of the circulation, vascular dilatation and contraction, or even necrosis, are caused by the action of various irritants, but inflammatory swellings are not generally absent.

2. *The Tongue.*—The tongue is drawn out of the mouth, stretched over a ring of cork, and fastened

to it by fine insect-pins, or hedgehog-spines which are then cut off short. Without further preparation it is generally too opaque for examination with strong lenses; a small piece is removed with fine scissors from the upper (originally the lower) surface. By avoiding the vessels that are visible, the escape of blood is prevented as much as possible; the blood is then washed away with salt-solution. If the field is clear, the cover-glass is adjusted; drying is prevented by dropping salt-solution upon the part, and the rest of the frog's body is wrapped in moist blotting-paper. The frog and the cork ring that supports the tongue are placed upon a glass plate, which is moved directly under the microscope. The observation can then be begun and continued for hours, and even for days. The tongue must not be stretched too much, lest an obstruction to the flow of blood result.

An object thus prepared is also perfectly suitable for strong lenses; the escape of the white and red blood-corpuscles can be demonstrated very well here, an artificial wound serving as an irritant to provoke inflammation.

3. *Mesentery.*—Use large male frogs (recognized by the glands on their thumbs), so as not to be embarrassed by the oviducts and ovaries. Incise the skin in the axillary line over the lower half of the trunk; the resulting hæmorrhage soon ceases, the incision is then carried through the muscles, and the abdominal cavity is opened for a distance of from ten to twenty millimetres. A coil of small intestine is carefully drawn out with blunt forceps, and is stretched over a cork ring and secured in the manner

described for the tongue. The mesentery must not be stretched too tightly, otherwise obstruction results; it is then covered with a piece of cover-glass, and presents a brilliant object for examination with the strongest objectives. The mesentery is kept from drying by the use of salt-solution, while the frog is enveloped in a damp covering. Here also the cork ring is not cemented to the glass plate, but it is much more convenient to leave it free. The more carefully the specimen is prepared, and the more all tension and other mechanical injury are avoided, so much the longer is it preserved, until exquisite inflammatory phenomena, arrangement of the white blood-corpuscles along the walls of the vessels, emigration, etc., take place. In the case of the prepared tongue, as well as in the mesentery, we can apply any desired irritant, or cause any injury to the tissue, as well as to the vessels.

The lungs and urinary bladder of the frog may be easily prepared for microscopical examination in a similar manner.

4. *Cornea.*—The long-lived cornea of the frog furnishes a good object for the observation of pathological processes, especially inflammation. The cornea, normal or inflamed (after cauterization, for example), is carefully excised and is placed upon the slide in the drop of aqueous humor which escapes at the time; the edge is then nicked in several places, so that the membrane may be spread out smoothly. The vital appearances of the cells, the wandering as well as the fixed, can be observed for several hours in the cornea after its removal.

## VI.

### THE EXAMINATION OF FLUIDS.

THE microscopic examination of fluids is extremely valuable for clinical and pathological purposes; a glance at the microscope frequently establishes the positive diagnosis of a disease which was previously obscure or misunderstood. The process offers but few technical difficulties, since it is simply a question of transferring a small drop of the fluid to the slide by means of a glass rod, and covering it with a cover-glass. The drop must not be so large that the fluid flows over the edge of the cover-glass, nor so deep that the glass floats upon it—this is all self-evident.

Our task now consists in examining the morphological elements contained in the fluid. These in a great many cases may be demonstrated at once with the naked eye, either in the form of a diffused cloudiness, or as coarser shreds or granular precipitates. These latter objects are, of course, first utilized for microscopical examinations, being collected with a small spoon, or a pair of forceps, and examined with powers of gradually increasing strength. The substance should always be first subjected to a careful microscopic examination by transmitted, as well as by direct light, a rule which, self-evident as it sounds, is however too often disregarded by the beginner.

If a very small number of formed elements are present that layer of the fluid is examined in which they are most numerous, generally the sediment, since the elements are in most cases specifically heavier than the fluid; only the fats float on the surface. It is less advisable to filter the fluid and to collect the residue from the filter, because in this way impurities can not be certainly avoided. In other cases the morphological elements are so abundant that an extremely thin layer must be used, in order to prevent them from lying in several strata, one above the other, and thus concealing one another. It is necessary to dilute very thick and pultaceous fluids in order to render the examination possible; serum, or as a rule, salt-solution, having a strength of three-fourths of one per cent., being used for this purpose.

*The Vital Properties of the Suspended Elements. Amœboid Movements.*—Aside from this last instance, we also encounter the elements in their normal menstruum, and can therefore be sure that they appear as little changed as possible, provided that all unfavorable influences are avoided. The various precautions must be especially observed if we intend to study the vital phenomena of the elements suspended in the fluid. The pressure of the cover-glass is next to be considered, since this may become considerable, not only by reason of the weight of the glass, but still more through the capillary attraction exerted in a thin layer of liquid; it must accordingly be supported, if necessary, by placing beneath it bits of glass, such as fragments of covers.

Furthermore, the fluid must be prevented from evaporating, and thus changing its concentration; this evil results very quickly at the edge of the specimen, but more slowly in the center, so much more slowly in fact the deeper the layer of fluid and the further we are removed from the edge. The evaporation may be reduced to a minimum by placing the preparation in some sort of moist cell. Take simply a wide glass tube, about two or three centimetres high, as a piece of a lamp-chimney; the interior of this is lined with a thick layer of moist blotting-paper, and the glass is placed over the specimen as it lies upon a broad slide; the opening is almost entirely closed above by the tube of the microscope.

It is better to examine the fluid in the form of a suspended drop, by using slides with cells that have a depth of one or two millimetres; the latter are easily constructed by cementing glass rings or borders to the slides, or they may be obtained from opticians under the name of "hollow-ground slides." If then the edge of the cell is oiled, and the drop of fluid is placed in the middle of the cover-glass, upon its under surface, a hermetically-sealed space is formed, within which no further evaporation occurs. This is the manner in which fluids must be examined in order to study the motor phenomena of the contained schizomycetes. If two tubes then open into the cell, we can study under the microscope the influence of gases upon the elements contained within the suspended drop (gas-cell). The so-called "amœboid" protoplasmic movements, as well as the processes of division in living cells may be observed in this way; during this examination all currents in the fluid itself must be avoided, lest a whirling about of the elements be mistaken for a change in their shapes. The colorless cells of the blood and lymph, pus- and mucus-corpuscles, many cells met with in exudations, and even tumor-cells, afford an opportunity for these highly interesting and engaging observations. He who intends to undertake such examinations must proceed ever with the most painstaking care, and must set to work critically, but above all with great patience. The movements are nearly always very slow, even when the warm slide is used; Stricker's model of the latter is to be recommended.

*The Form of the Elements.*—We are, however, mostly concerned with elements the form of which is perfectly constant, so that we have simply the task of studying these accurately. To this end it is necessary to view the body in question from all sides; for it is quite clear that a circular figure, for example, which is observed under the microscope may represent either a disk, a sphere, a cylinder, or a cone; even an ellipsoid, an oval, or a still more irregularly shaped body, may, under certain circumstances, appear in the microscopical image as a circle. We are aided here first of all by the micrometer-screw, since we obtain by using it the outline of the observed object at different focuses, and thus form a combined stereoscopic image; also by the passive movements which we can cause the object to make by allowing it to revolve around its different axes; the simplest way to effect this is by exciting a current in the fluid, either by means of a bit of absorbent blotting-paper placed at the edge of the cover-glass, or by pressing upon the cover with a needle. The beginner will sometimes during the course of these manipulations, not only turn and roll the body in question, but will cause it to disappear entirely from the field of vision; however, he soon acquires the necessary delicacy in graduating the pressure, and can then discriminate sharply concerning the forms of elements as seen from different aspects and thus easily determine their stereometrical figures.

*The Examination of Tissue-fluid, etc.*—In examining fresh organs it is often of great importance to promptly observe the isolated elements (cells, etc.) of the same; in many cases this is accomplished in an extremely simple manner by examining the tissue-fluid which has been wiped away. For this purpose a freshly cut surface is always exposed, and this is scraped with the blade of a scalpel. In proportion to the stability of the elements on the one hand and the firmness of their cohesion (or that of the cement-substance) on the other, we succeed, by employing firmer or lighter pressure, in isolating in this simple and very convenient way the elements of most parenchymatous organs, or at least some of them. We must of course always bear in mind the narrow limits of the method; but much time may be saved in this manner, since it is often by no means necessary in order to answer certain questions to examine accurate sections of the organ under consideration, that is, if the isolated elements are sufficiently characteristic.

The tissue-fluid that has been scraped off must usually be diluted before it is examined microscopically; salt-solution is used for this purpose as a rule. A fine glass capillary tube can be inserted into many soft tissues and the tissue-fluid with its suspended elements be drawn into it; E. Neumann employs this method particularly in examining the lymphoid cells of marrow. The elements are thus always obtained in their natural menstruum. It is possible to isolate the elements of soft tissues in a very simple manner by lightly teasing with needles; the bit of tissue is rapidly torn into small pieces in a drop of salt-solution. The fluid is thus filled with the separate cellular elements that are removed from the fragments; these elements are examined as they float freely in the fluid, and so are all of the bits of tissue which have become sufficiently transparent, at least at their edges. In the case of fibrous tissues, as muscles and nerves, the elements are isolated in the direction of their vertical axes by careful teasing with needles.

*The Examination of Micro-organisms.*—On account of the great and ever-increasing importance, as well as the peculiar character of the subject, it is advisable to discuss separately the examination of fluids for micro-organisms, especially for schizomycetes.

First of all it is evident that in these investigations every impurity must be rigidly excluded; care must be taken in obtaining the fluid to insure absolute cleanliness of the vessels, canulæ, etc. Furthermore, the objects must always be examined when they are perfectly fresh; micro-organisms can develop in great numbers within a few hours since their germs are everywhere diffused. The latter are present on the sides of every vessel, no matter how clean it is, and on every wiping-cloth; in smaller numbers also in the atmosphere, especially in inhabited rooms (hospital-wards, laboratories, etc.), so that it requires special precautions, such as the prolonged heating of all vessels to above 100° C., in order to collect and to preserve fluids without contamination with the accidental germs of minute organisms. The *generatio equivoca* of cleft-fungi, the doctrine of which was repeatedly revived a few years ago, always resulted from the neglect of some one of the necessary precautions.

The fluid is always examined in a perfectly fresh condition, that is immediately after its removal from the living or dead body; in the latter case we must ever bear in mind the possibility of a post-mortem origin. A method introduced by R. Koch is strongly recommended: the sample of fluid is removed and transferred to the slide by means of a platinum wire cemented to a glass rod, since the former can be very easily and surely cleaned immediately before and after using by heating it.

The fluid is first examined directly, without the addition of any reagent, because in this way we are quite sure that any organisms which may be found really belong to the fluid itself. In many cases the organisms are recognized by their active movements. But it is necessary to use great care in this connection, because small bodies suspended in liquids nearly always show, under certain circumstances, a very active movement—the Brownian molecular motion.

We do not as a rule form a proper conception of the energy of these movements, which are for the most part due to currents caused by evaporation. In order to obtain an idea of them sprinkle some finely powdered carmine into a drop of water, and examine this with a high power; you will be extremely surprised at first at the rapidity and apparent spontaneity of the passive movements of the carmine granules. Before, then, an opinion is hazarded concerning the " spontaneous movements " of granules which you are inclined to call micro-organisms, it is strongly advisable to familiarize yourself perfectly with Brownian molecular motion. Even if you think that you have to do with vital movements, you must always prove this by showing that the motion ceases when such agencies are introduced as are incompatible with the life of the organisms, such as heat, or treatment with strong acids and alkalies.

Most of the micro-organisms that are of interest to us (especially mold-fungi, yeast- or sprouting-fungi, and cleft-fungi or schizomycetes) resist these reagents strongly; the spirochætes only, which occur in the blood during recurrent fever, form an exception since they quickly perish in all the different reagents, even in distilled water. This capacity of resistance of the microbes can also be utilized for their diagnosis, because granules of protoplasm, for instance, dissolve in strong acids and alkalies while schizomycetes remain unchanged. If the latter are grouped in so-called colonies (as the zoögloe-masses) they often stand out clearly after treatment with strong acetic acid or liquor sodæ, since the cellular elements and other granular matters, which previously concealed the colonies, are entirely cleared up. Their disposition in chains, or the characteristic form of the separate individuals (rod-shaped, oval, etc.), often renders the diagnosis of micro-organisms possible without further difficulty. We must always guard against mistakes; granular unorganized precipitates may be taken for micrococci and minute crystals for bacilli, and even very small fat-granules may cause a careless observer to err.

*Koch's Method of Staining Dried Preparations.*—Under some circumstances it is certainly by no means easy, and is often quite impossible, to arrive at a positive opinion regarding the significance of minute granules that are contained in any fluid by a simple examination, and by the application of the ordinary micro-chemical reactions. In these, as well as in all difficult cases in general, in fact in every instance where it is desirable to make permanent preparations, the method of drying and staining is employed, which originated essentially with Koch and Ehrlich.

This depends upon these two facts: 1. If a thin layer of fluid is dried rapidly, the forms of the cellular elements and schizomycetes are not materially altered; 2, the schizomycetes are marked by their great affinity for the basic aniline dyes, and may in this way be distinguished from other granular bodies. It should always be observed that not only the schizomycetes, but other bodies also (as cell-nuclei and their fragments and certain protoplasmic granules) show the same affinity for the dyes in question, so that in using this method it is necessary to criticise strictly the value of the objects found. It is also probable that there may be varieties of schizomycetes which do not possess this affinity; the forms thus far known all show a very strong capacity for dyeing, but many, however, stain only under certain fixed conditions. The process is as follows: The fluid is spread out in the thinnest possible layer upon the cover-glass or slide, either by separating the drop into a very thin layer by means of a needle or platinum wire, or by placing another cover-glass upon it and again removing the same. The beginner easily fails by making the layers too thick; they must be as thin as those used for the examination of the blood. Then the fluid is dried in the air, and is exposed for a few minutes to a temperature of 120° C.; it is enough to pass the glass with the dried liquid carefully through the flame of a gas-burner three times, with a motion about as rapid as that with which one cuts bread (Koch). The necessary experience in warming the specimen sufficiently, without overheating it, is soon attained. Warming is especially required for fluids that are rich in albumin, its principal purpose being to transform the albumin into an insoluble variety; it must not be prolonged above five or ten minutes when schizomycetes are present, else their staining power will be impaired. After warming* the preparation is stained. A drop of a strong solution of gentian-violet, methyl-blue, fuchsin, or Bis-

---

* Heating may be dispensed with in the case of non-albuminous fluids; layers of strongly albuminous fluid, on the contrary, if they are simply dried and are then treated with staining-solutions, are apt to swell readily and to partially dissolve; hence with these the heating or coagulation of the albumin precedes.

marck-brown, in short of any basic aniline dye, is poured upon it and is left standing for a little while (from one to several minutes), after which it is washed off with distilled water; a brown, blue, or red cloud is then observed at once upon the glass. The examination may be undertaken directly, by placing the cover-glass, with the dried and stained layer of fluid upon its under side, on the slide with a drop of distilled water. The water adhering to the upper surface is easily removed by sucking or blowing it with a glass pipette. If the under surface of the cover is again dried (and this is also accomplished most quickly by blowing upon it through a tube), the specimen may be at once permanently mounted in a drop of Canada balsam, prepared with chloroform. If a longer time (several minutes) is desired for staining, the coloring of the film of fluid that has been dried on the cover may be effected in a watch-glass. Most of the other granular elements are also stained, but methyl-blue has the advantage, according to Ehrlich, of not overstaining even after acting for hours.

The cellular elements may generally be preserved perfectly well in their original forms in such a preparation; the changes of shape in some of these, caused by the spreading out of the fluid (comet-shaped figures), are very readily recognized as such. The nuclei and also the schizomycetes are stained with special intensity, so that in this way they are brought into view in a striking manner. Ehrlich has recently recommended very highly for staining schizomycetes, methyl-blue in a concentrated watery solution that must act for half an hour and longer. In the author's experience the before-mentioned dyes, especially gentian-violet, have given about the same results as methyl-blue.

Every one who becomes acquainted with this simple method must adopt the opinion that these are undoubtedly the best ways hitherto known of demonstrating micro-organisms in fluids (that is, with the modification proposed by Gram. Compare p. 75). In examining specimens thus prepared with strong immersion-systems and with an open condenser, we recognize at once the sharply-defined, deeply-stained micro-organisms in their characteristic forms and groupings, and learn very soon to diagnosticate as such certain impurities or precipitates, and to distinguish them from schizomycetes. In mucous fluids principally, such as synovia, we have to contend with precipitates, since the mucus always takes a pretty deep color with these dyes; nevertheless it only requires slight practice in order to show properly the irregular, granular, and thready masses. By treating the stained specimens for a short time with a dilute solution of iodine and iodide of potassium, the color of the schizomycetes usually appears more intense; or the nuclei may be subsequently decolorized by treatment with alcohol, and separate staining of the parasites may be produced (Gram).

Finally, it should be observed that *tubercle-bacilli are never stained by this method;* they differ in this respect from all the other known forms of schizomycetes. Baumgarten has even proposed to make use of their negative peculiarity, or failure to stain, for their rapid recognition in sputa.

Tissue-fluid can be examined for schizomycetes, after being dried and stained, precisely like ordinary fluids. If a fresh, clean-cut surface of an organ be stroked with a heated platinum wire a sufficient quantity of liquid is generally obtained, which is then rubbed directly over the cover-glass with the wire. This process is very valuable as furnishing the most convenient and rapid demonstration of the micrococci of pneumonia.

He who desires to pursue original investigations will naturally not be satisfied with the simple demonstration of micro-organisms, but will be obliged to study their peculiarities more closely. The method of drying upon the cover-glass is very convenient for this purpose; fifty or more dry preparations can be very easily made from a single fluid or from a cut surface of an organ, and can be kept unstained in a small dish as long as is desirable. In this way a large number of nearly identical cover-glass preparations are obtained, which can be studied in various ways by the aid of chemical reagents and different stainings.

It is quite clear that all schizomycetes do not react in the same manner with all the basic aniline dyes; the most important differences, so far as is yet known, are those displayed by the encapsulated micrococci of pneumonia. However, we may assume that, with continued investigations, other schizomycetes will be found to manifest specific reactions of this character.

1. BLOOD.—The examination of blood is easily effected, in accordance with the principles already stated. A drop of blood, obtained either by copious bleeding or by the prick of a needle, is taken up neatly and covered; great care must be employed to have the layer extremely thin, so that never more than a single stratum of blood-corpuscles may be present. When the drop is to be taken from a punctured wound, the skin in its neighborhood must be carefully cleaned and dried, and the needle that is to be employed must be heated just before using; in spite of these precautions we should be prepared to meet with certain impurities, though only such as epidermal scales, etc. It is better to wipe away carefully the drop which first wells up, to allow a new one to appear, and then to take up a small portion of this on a cover-glass that is held over it; the cover with the adherent drop (which should be at most not larger than the head of a small pin) is then placed gently on the slide, so that the blood is spread out in a very thin layer between the two glasses.

If the layer has been made sufficiently thin, we at once recognize between the discoid red blood-corpuscles the clear transparent plasma and the white cells, which, as is known, exist in small numbers in normal blood. There are also found in normal blood variable numbers of small, irregular granules or particles, which have been described as elementary granules or products of degeneration. There is still a difference of opinion as to their significance. It is probable that elements possessing quite different degrees of importance are hidden away among them;

the "blood-plates" recently described by Bizzozero, which, according to this author bear a close relation to coagulation, have always been regarded hitherto as indifferent bodies resulting from decomposition. Hayem describes them as hæmatoblasts, or elements out of which the red blood-corpuscles are to be formed; it is highly probable that he is wrong. The number and size of these little bodies rarely vary; whether they possess any pathological significance or not is still uncertain. They were once regarded by Lostorfer and Stricker as characteristic elements of syphilitic blood, and were called "syphilis-corpuscles"; if this view were correct every man would be syphilitic.

As regards the action of the ordinary reagents, it is well known that distilled water, as well as acids and alkalies, causes the red corpuscles to swell and become pale; the hæmoglobin is rapidly discharged so that the corpuscles almost entirely disappear. In order to preserve them as far as possible in their natural form, solutions of a certain strength (so-called indifferent fluids) must be employed: for example, a solution of sodium chloride from three-quarters to one per cent.; concentrated salt-solutions always preserve the blood-corpuscles, but they occasion essential changes in their form and size—that is, they cause them to shrivel up.

It is always advisable in examining abnormal blood to observe it directly and when undiluted. The following are the most common changes in the blood:

(*a*) *Diminution of the Number of Red Blood-cor-*

*puscles in Anæmia.*—It is easy after a little experience to establish at once, without the aid of additional apparatus, the existence of the more marked degrees of this change, by comparing a specimen of the pathological with one of normal blood prepared in the same manner. If more exact determinations are desired, a blood-counting apparatus is required; the instrument most worthy of recommendation is perhaps the one devised by Thoma,* which is constructed by the optician Zeiss in Jena. The exact directions for using this apparatus will be found in the passage cited. While the normal number of red blood-corpuscles is about five million to the cubic millimetre, it may in severe cases of anæmia fall to five hundred or even to one hundred and forty-three thousand (Quincke); it is evident that advanced grades will have already been clearly recognized without special counting.

(*b*) *Change in the Size and Shape of the Red Blood-corpuscles, or Poikilocytosis. Nucleated Red Blood-corpuscles.*—The red corpuscles, as is known, all possess normally the same characteristic discoid shape, with a depression on both sides; the center is thinner, and is therefore less deeply colored than the edge. The size of the normal red blood-corpuscles also varies within relatively narrow limits, as any one can readily prove to himself. There appear in most cases of anæmia, but with especial regularity in so-called idiopathic pernicious anæmia, beside normal corpuscles numerous irregular and, as a rule, smaller

---

* Lyon and Thoma, "Ueber die Methode der Blutkörperzählung," "Virch. Arch.," Bd. 84, S. 31.

bodies, containing hæmoglobin, called microcytes, and sometimes also others which exceed in size the normal red disks (megaloblasts of Ehrlich). Nucleated blood-corpuscles are observed (though rarely) in the direct examination of anæmic blood. Ehrlich has found, by examining dried and stained preparations of blood (obtained according to the method described on page 122, except that the warming occupies more time), that nucleated corpuscles may be demonstrated in all severe cases of anæmia, whether of traumatic or idiopathic origin. He showed this point of difference, that in traumatic or secondary anæmia nucleated blood-corpuscles are found of the same size as the normal red disks (normoblasts), while the large forms, or megaloblasts, are characteristic of idiopathic anæmia.

As regards the microcytes and poikilocytes, it should be stated that they very probably represent a degeneration-product, or a form of disintegration of the normal corpuscles; many analogous forms can be found in the blood of a cadaver when examined about twenty-four hours after death, and under certain conditions they may be seen to develop in the specimen of blood under one's very eyes (Vulpian). In every case in which we desire to observe these objects the blood must be examined in a state as little changed as possible. Certain forms of microcytes are, however, to be regarded as artificial products.

(c) *Increase in the Number of the White Blood-corpuscles. Leucocytosis and Leucæmia. Changes in the Granular Protoplasm.*—In many affections, especially in febrile conditions, the white blood-corpuscles are increased absolutely and relatively as compared with the red (leucocytosis of Virchow). The proportion of the white to the red corpuscles is normally one to three hundred, or even less; we can easily learn from a direct examination of the blood (which should be observed in this case also in its natural state) to estimate pretty accurately the relative increase of the white cells. In leucocytosis, a condition that may again retrograde, the white corpuscles are increased to one in fifty, and even more; in leucæmia, which is, as a rule, a persistent and necessarily fatal disease, the proportion increases so much in the most serious cases that the white corpuscles exceed in number the red. At the same time the absolute number of the red disks is very considerably diminished; the counting-apparatus is necessary for the precise determination of these proportions.

The condition of the granular protoplasm within the leucocytes is a subject of extreme interest; this has been studied of late by Ehrlich.\* Ehrlich prepares dried specimens of blood in very thin layers, and heats these for some time at a temperature of about 120° C. If different staining-solutions are now allowed to act upon these preparations, constant variations in the coloring of the protoplasmic granules within the leucocytes are obtained, which are of great physiological, as well as diagnostic, significance. He distinguishes in this manner five different kinds of granules, from the alpha to the epsilon

---

\* Ehrlich's statements with reference to his methods of staining and their results are scattered throughout several dissertations, written by his pupils, as well as in a number of minor articles in different places—a very injudicious way of publishing.

variety; the *a* granules, also called *eosinophil* granules, are characterized by their property of staining deeply with acid dyes, such as eosin. These eosinophil granules are present in only a very few normal human white corpuscles, and it is possible to distinguish a beginning leucæmia from an ordinary leucocytosis by the presence of numerous eosinophil cells. The demonstration of these cells is, according to Ehrlich, very simple; a dried and warmed blood-preparation having been rapidly stained with a drop of a solution of eosin in glycerin, is washed in water, and is then dried and mounted in Canada balsam. If the eosinophil cells are increased in number, they will be seen at once as red bodies.

(*d*) *Other Cell-Elements which appear in the Blood. Worms and Schizomycetes.*—During typhoid fever there are found in the blood large cells, which contain in their interior several red blood-corpuscles (Eichhorst). It is quite probable that these come from the spleen, since at autopsies similar bodies are found regularly in the splenic tumor of typhoid. Flat cells loaded with fat-drops are frequently observed in the blood in acute infectious diseases, especially in recurrent fever; they are regarded as endothelial cells from the walls of the vessels. Cells containing granules and flakes of black pigment, as well as free masses of the same, appear in the blood in severe malarial poisoning (melanæmia). Tumor-cells, which circulate in the blood in cases of malignant metastatic growths, we can scarcely expect to find in examining the blood obtained from the capillaries by a needle-puncture or cupping-glass; these cells are only characteristic, as a rule, when they reach considerable dimensions, so that they can not pass through the narrow capillaries. Of the animal parasites, *Filaria sanguinis hominis* and *Distoma hæmatobium* appear in the blood of man, but both are found only in tropical, or sub-tropical, countries.

Schizomycetes have hitherto been regularly discovered in human blood only in two affections, the *Bacillus anthracis* in anthrax (Davaine), and the *Spirochæta Obermeyeri* in recurrent fever.* The blood is examined directly in a very thin layer, without the addition of any reagent, or dry preparations may be heated in the manner already described, and stained with gentian-violet, methyl-blue, etc. The anthrax-bacilli are slender, motionless rods which resist most reagents; the spirilli of recurrent fever, on the contrary, move very actively, and are easily destroyed by most fluids that are added to them, even by distilled water. The presence of the spirilli is, as is known, limited to the febrile stage of the disease; they are found only very rarely for a short time after the attack. They are, on the other hand, never absent during the progress of the fever, so that they may be regarded as a sure diagnostic criterion of this affection. However, they are sometimes present only in small numbers, even in rather severe cases, and may perhaps be overlooked on a cursory examination. It is advisable in these instances, if it appears to be of sufficient diagnostic importance, to remove several grammes of blood by means of a cupping-glass, and

* Tubercle-bacilli have as yet been found by Weichselbaum only in the blood of a subject after death from general miliary tuberculosis.

to allow it to coagulate. The spirilli are then apt to gradually collect at the edges of the clot, to the number of twenty or more, and are even rolled into a knot, or united together in the form of a rat-king (Rattenkönig*). Since, after being removed from the body,† they continue their active lashing movements for hours, or even days (although kept at the temperature of the room), when collected in groups they cause violent currents in the fluid, so that attention is often directed to them even when a low power is used. The spirilli may be very easily stained in dry preparations with the different aniline dyes, as gentian-violet. The history of their development is still unknown; whether the supposed movable corpuscles and double granules, which have been seen in the blood in recurrent fever, as well as in other infectious diseases, and sometimes even in normal blood, and have been described as "micrococci," or spores, are really to be regarded as such, is still doubtful.

In every case the reports concerning the occurrence of micro-organisms in normal blood are thoroughly untrustworthy; most of the communications also concerning the so-called micrococci, monads, or rods, seen in the blood during various infectious diseases, such as diphtheria, hospital gangrene, erysipelas, etc., as well as in intermittent fever, are not sufficiently supported. Even in pyæmia and in ulcerative endocarditis no organisms can usually be demonstrated by an examination of the blood of the

* This expression signifies several rats, which have been attached together by their tails.—TRANS.

† It has been recently stated that they are even observed to increase in number outside of the body (Albrecht).

living, while after death the capillaries in many places are found to be plugged with micrococci. It is not unlikely that in such cases the organisms enter the blood in batches, possibly only in small numbers, and that they then soon lodge in the capillaries, in which, under certain conditions, they may rapidly multiply. It should be remarked also that, although micro-organisms are often observed in pus-corpuscles, they have, as yet, never been certainly found in the interior of the human white blood-corpuscles, although the latter are accustomed to appropriate with avidity other finely granular matters. Probably the future will bring further discoveries in this direction; it is quite possible that within the corpuscle, which we have hitherto been forced to describe in a loose way as a product of degeneration, there lie concealed important forms, which we shall one day learn to differentiate.

(*e*) *The Examination of Blood-Stains. Hæmin-Crystals. Hæmatoidin.*—Blood, when dried upon wood, linen, metallic instruments, etc., is frequently the subject of medico-legal examinations. It is very often possible, after soaking in proper fluids, especially in salt solution (eight tenths of one per cent.) and liquor potassæ (thirty-three per cent.), to isolate blood-corpuscles from such stains, and even their shape and size are to some extent preserved.

Since now the blood of man and of mammalians is, as is well known, characterized by the circular form of the red blood-corpuscles, while the other vertebrates have oval corpuscles, we can generally decide with perfect certainty, even in the case of dried blood, whether it comes from a mammalian (including man) or from some bird, etc. But beyond this our art does not extend; it is impossible to positively identify given blood-corpuscles as human. Most mammalians, however, possess rather smaller corpuscles than man; those from the blood of the sheep and goat are on an average only a little over half as large as human corpuscles, but other animals, as the dog for example, resemble man very closely in this respect. It is better in every instance, when deciding upon blood-stains, merely to state in a given case that the blood came from a mammalian; it is impossible to assign a positive value to the relations of size, because of the differences in the mode of drying, and of swelling when fluid is added, which vary with the age of the spot, etc. The so-called Teichmann's hæmin-crystals are also obtained from dried blood. These consist of hydrochlorate of hæmatin, and are prepared in the following manner: To a small quantity of dried blood placed upon a slide (for example, a thread dipped in blood) add a few drops of glacial acetic acid and a particle of salt; then warm the slide gradually until bubbles begin to form. There will be observed, forming around the thread, a large number of dark brown rhombic crystals, which are perfectly insoluble in water, and show exquisite double refraction. This test furnishes positive results even in the case of quite old blood-spots; it is self-evident that this process can be applied to every blood-stained object. It is frequently necessary to first soak out the suspicious spot in water, and then to apply further tests to the watery extract.

From these artificial products are to be distinguished the hæmatoidin crystals, which also occur in a rhombic shape in old blood-extravasations, in the corpora lutea of the ovaries, etc., partly free and partly inclosed in the interior of cells. They are of a bright ruby or orange-red color, but contain no iron; they are soluble in chloroform and resemble bilirubin very closely, in fact they are regarded by many as identical with bilirubin.

Hæmoglobin-crystals, so far as is now known, do not appear normally in man, while these and other bodies have been found in large numbers in the puerperal uterus of the Guinea-pig; they may be prepared in different ways, for instance, by the action of a concentrated solution of pyrogallic acid upon blood that has been previously diluted with distilled water.

2. SPUTA. — The microscopical examination of sputa is of extreme diagnostic importance, and is therefore very frequently undertaken. The technique of this examination is quite simple: it is first determined by gross inspection what substances can be distinguished in the sputum, since the latter always presents a number of different kinds of material which arise from various sources. Particular attention must be paid to opaque white, or grayish-white, plugs, which are best seen if the sputum is spread in a thin layer upon a polished black porcelain plate; in these plugs we generally find at once elastic fibers, which should be preserved, since they are to be regarded as cast-off pieces from the wall of a cavity. The so-called "asthma-crystals" of Leyden are also found as a rule in the interior of greenish-

white miliary plugs, which may be distinguished, even by the naked eye, in the otherwise clear sputum. If also we have occasion to look for echinococci, or for other rarer additions, in short in every case, a careful macroscopical examination should first be made; a neglect of this caution often leads to errors and negative results. Microscopical preparations are then made of all these different portions, which are frequently arranged in layers one above the other, by transferring a bit of mucoid material to a slide, by means of a needle and spatula, and covering it with a cover-glass. It is seldom necessary to employ a diluting fluid, such as salt-solution or distilled water. In examining microscopically, begin with a low power (from fifty to eighty), and only change this for stronger lenses after you have studied the entire specimen with the former. The elastic fibers are usually recognizable to some extent, even with a weak lens, either directly, or by the dark, friable material in which they are imbedded; since the low power furnishes a correspondingly larger field of vision, there are more chances of finding them with this than with a high power. Besides, the former is not so much affected by slight differences of focus, so that a number of superposed layers in the specimen may be examined simultaneously.

Those portions of the preparation which are recognized by the low power as being peculiar, are then analyzed more closely with the high objective; then only is the microscopical diagnosis established.

(*a*) *Oral Fluids.*—Impurities of the most varied description are almost always encountered during the examination of sputa. We need never expect to obtain the contents of the bronchi in an unmixed state; the secretions of the mouth, salivary glands and throat at least are mingled with it. We must accordingly learn to recognize these perfectly. Large numbers of epithelial cells from the mucous membrane of the mouth and throat are always found normally in the oral fluids. You will very soon learn to recognize these large, irregular plates which, when treated with acids or alkalies, swell up to form globular vesicles, and are generally filled with numerous schizomycetes. In catarrhal conditions of the mucous membrane of the mouth and throat there also appear in the oral fluid living epithelia, which throw out knobbed processes and execute feeble amœboid movements.

In oral catarrh the superficial layers of epithelial cells are often cast off *en masse*, and in the coating on the tongues of individuals thus affected we find in very large numbers the goblet-shaped tips of the filiform papillæ, which consist of resistant, horny epithelial cells, closely adherent to one another.

We find also many round cells, mucous or salivary corpuscles, which come from the mucous or salivary glands (especially from the submaxillary and sublingual), and to a less extent also from the deeper layers of the stratified pavement epithelium. These are originally small amœboid cells, analogous to lymph-corpuscles, but under the influence of the dilute parotid saliva they become changed, assuming a globular shape, while a limiting membrane stands out in contrast with the clear contents; the latter includes one or two round nuclei, besides a large number of fine granules, which are always seen in active dancing motion—molecular motion in salivary corpuscles. These granules are not, as has been supposed, parasitic organisms, at least they never stain in aniline dyes; whether their active movement represents a vital phenomenon or not, somewhat analogous to the protoplasmic currents in the cells of plants, has not been determined.

But, besides these, numbers of micro-organisms of different sorts roll about freely in the fluids of the mouth, such as stiff, long threads of leptothrix of varying breadths; globular cocci, of different sizes, frequently arranged in the form of chains or compact groups, and not seldom bacilli; and very beautiful forms of spirochætes, which in their shape and serpentine movement closely resemble the spirilli of recurrent fever, except that they usually attain much larger dimensions than the latter. The less careful a person is about cleansing his mouth, the larger are apt to be the masses of micro-organisms which collect there; but they are never entirely absent, even when the greatest cleanliness is observed, for the germs of schizomycetes, that are always contained in the inspired air, invariably find in the fluids of the oral cavity the most favorable nidus. One species of these was regarded as the cause of dental caries, but this is probably erroneous (W. Miller). By inoculating animals it is found that there are a great many pathogenic cleft-fungi in the normal human saliva.

Budding fungi are only occasionally present in large numbers in the oral fluids, but a species of thready fungus, the parasite of aphthæ, or *oidium albicans*, is frequently found; this occurs, as is known, most often in bottle-fed children, and in adults whose systems are greatly reduced, as in the phthisical. This fungus consists of branched or limb-like threads with oval spores, which form a more or less thick mycelium between the epithelial cells of the superficial layers and upon the surface itself.

There often appear in the saliva, in addition to these elements, various remnants of vegetable food. For several hours after a meal many persons retain in their mouths, especially between the teeth or in carious cavities of the same, microscopical samples of their repast, such as often astonish and perplex the beginner when examining the expectoration.

The mucous fluid contained within the nasal cavity, with which blood is frequently mixed, may also occur in the expectoration as an impurity, as well as the contents of abscesses that open into the cavity of the mouth and throat (dental, alveolar, tonsillar, and retropharyngeal abscesses). Special attention should be called in this connection to the small concrements, sometimes exceeding a pea in size, which often form in the crypts of the tonsils, through the incrustation of the retained secretion with lime, and are occasionally dislodged during a fit of coughing. The patient, and often the physician also, are greatly disturbed by the appearance of this supposed pulmonary calculus, but the correct diagnosis is very easily made on microscopical examinations. The lime is dissolved by the addition of dilute hydrochloric acid, and there remain then the large, horny, epithelial cells (which are sometimes disposed in the form of concentric balls) and numerous micro-organisms.

(b) *Products of the Respiratory Mucous Membrane.* —After you have made yourself perfectly familiar with all these elements which appear accidentally in the expectoration, you can proceed with profit to the microscopical examination of the latter. The chief component of the expectoration is the excretion of the inflamed mucous membrane of the respiratory tract, a fluid containing more or less numerous round cells. The secretion of the mucous membrane of the throat and upper part of the larynx, as far as this is lined with stratified pavement epithelium, is also exceedingly rich in cast-off cells; in the upper part of the respiratory tract, on the contrary, as well as in the lower portion of the larynx, trachea and bronchi, and at every point where the membrane is covered with long, cylindrical ciliated epithelium, desquamation occurs only in exceptional cases. It is rare to meet with ciliated cells, or their remains, during the examination of sputa.

The vitreous, transparent portion of sputa contains but few formed elements; the more numerous the latter, so much the more turbid and opaque is the sputum.

Round cells occur most frequently in purulent expectoration, in which case there are generally numbers of fine fat-granules in the protoplasm of the cells, which increase their opacity, and give them a yellowish tint. The round cells are, in most instances, dead and motionless, and consist of protoplasm containing dark granules and one or more nuclei; a special wall is not usually present, but the edge of the cell presents fine indentations which look as if they had been gnawed out. The protoplasm is usually so extremely granular that the nucleus (or nuclei) is concealed; the latter comes out more clearly after being treated with acetic acid, while the granules of protoplasm for the most part disappear. The majority of the round cells that are found in the sputum are about the size of white blood-corpuscles, or a little larger. It may, in fact, be assumed that a large part of them are to be regarded at once as emigrated white blood-corpuscles, while another part may originate from the mucous glands, or from the inflamed mucous membrane.

But in addition to these we find not infrequently larger round cells of an epithelial type, which are characterized by thin, clear, somewhat circular outlines and single vesicular nuclei; that is, the nucleus is bordered by a dark, sharply-drawn line, while its interior is clear and contains one or more nucleoli. If these epithelioid cells are stained (which is best accomplished by treating the dried and heated preparation with basic aniline dyes) the nuclei do not take such a dark and even shade as those of the ordinary small round cells, but a difference may always be recognized between the deeply-stained periphery and the clear center. A few fat-granules are also found now and then in the protoplasm of these larger cells, and often also black pigment, in the shape of minute granules or irregular scales—so-called lung-pigment; blackish, or coarsely-gray spots and streaks are frequently observed, even on macroscopic examination. By microscopical analysis it is proved that the black staining is hardly due to the abundance of pigmented cells, and that free pigment is usually present in only a small amount. These dark granules and scales in the cells are all to be regarded as inhaled coal-dust that has fallen upon the surface of the respiratory mucous membrane, has been retained there, and has been subsequently taken up into the interiors of the amœboid cells. These pigmented cells are almost never absent from the secretion of the throat and respiratory tract in the case of individuals who have ample opportunity to inhale coal-dust, such as miners and inveterate smokers; but every man who, under our conditions of civilization, lives for the most part in closed and heated apartments, or even in the open air exposed to the influence of sooty chimneys, has an opportunity with every breath that he draws to introduce finely-divided coal-dust into his respiratory organs. The more mucus there is secreted on the surface of the respiratory tract, the more coal-dust there is retained; a portion of it makes its way into the interstitial tissue of the lungs and into the lymph-glands, through the aid of the amœboid cells, while another part is cast off again with the sputum.

Genuine pigment, however, or that which arises within the organism itself, appears very seldom; it is at once recognized by its brownish (not black) color and it points either to a previous hæmorrhage, or to an obstruction in the venous circulation of the lungs, with the escape of red blood-corpuscles and the transformation of the same into pigment (brown induration of the lungs). Hæmatoidin-crystals are sometimes found also. In the larger epithelioid cells of the sputum which have been referred to there are often seen collections of homogeneous, glistening granules, which, purely on account of their external resemblance, are described as "myelin-granules"; nothing certain is known regarding their nature and significance. In Buhl's well-known demonstrations* particular stress was laid upon these larger round cells in sputum; they were considered to be desquamated epithelial cells from the alveoli of the lungs, and because of their occurrence in great numbers in the expectoration the diagnosis of "desquamative pneumonia," that is of incipient phthisis, was made. These views have been shown to be untenable; large numbers of these cells, even gathered together in masses, and filled with granules of "myelin," fat and pigment, have been found in the morning sputum of entirely healthy men, and in simple bronchial catarrh.† It is likewise extremely doubtful if they were really regarded in all cases as cast-off epithelial cells of the pulmonary alveoli; on the contrary it has become probable as the result of many examinations that under some circumstances ordinary migratory cells (leucocytes, etc.) may be transformed into such epithelioid elements.

Among other cellular forms, we occasionally encounter cells filled with fat-granules (in sub-acute pneumonic processes), and rarely giant-cells (in tubercular phthisis).

* Buhl, "Lungenentzündung, Schwindsucht, und Tuberculose," München, 1872.

† Compare Guttmann und Smidt, "Zeitschr. für klin. Med.," Bd. 3. Panizza, "Deutsch. Arch. für klin. Med.," 1881. Bizzozero, "Centralbl. für klin. Med.," 1881, S. 529.

(c) *Elastic Fibers. Fibrinous Exudation. Asthma-Crystals.*—The finding of elastic fibers in the sputum is naturally of very great importance, since through these we infer at once a degenerative process in the interior of the lungs. As already stated, we find elastic fibers in the opaque plugs before described; it is frequently possible, by the addition of strong acetic acid, to render even very thick portions of these plugs quite transparent, so that the elastic fibers within them, which are known to resist the acid, appear most beautifully. The beginner sometimes regards as suspicious small bits of cotton-fibers, etc., which also resist acetic acid; needle-shaped crystals of fatty acid which often appear in great quantities in sputa during putrid bronchitis, pulmonary gangrene and similar affections, may possibly, by reason of their long, wavy shapes, give rise to error. These melt when heated gently and are transformed into small fat-granules. We should lay it down as a rule to make the diagnosis of elastic fibers only when several of these are united and, by their characteristic wavy course, make it clear that the framework of an alveolar wall is present.

We frequently meet with larger coherent pieces of the framework of the lung, including several alveoli; the more numerous and the larger these microscopical fragments of lung are, so much the more serious and rapid must we consider the destructive process.

It should be remarked also that in pulmonary gangrene we sometimes fail to find any elastic fibers in such expectorated fragments of lung as are recognizable under the microscope; these are gradually dissolved in the putrid fluids. Traube, who called attention to this condition, recognized in this a point of differential diagnosis from pulmonary abscess, in which affection the elastic fibers are preserved longer. Even in ordinary caseation of the lungs the elastic fibers disappear, though very gradually.

Fibrinous masses also appear in sputa under certain conditions—even coarse fibrinous molds of the bronchi in croupous bronchitis, and of the bronchioles in acute pneumonia (Remak). The branched, dichotomous masses of fibrin, which are already recognizable by the naked eye, show under the microscope the familiar appearance of a net-work of delicate fibers, which swell up and disappear in acetic acid, in which are imbedded numerous round cells and quantities of micrococci. A brief reference has already been made to asthma-crystals; these are pointed octahedra, occurring in large numbers (within the plugs already described) in the sputum of asthmatics during an attack, while they are generally absent in the interval. They are occasionally found also in non-asthmatic sputum, so that they are not pathognomonic of this disease. They are perfectly analogous to the crystals which have been found in the semen, in the marrow of bones, in the blood (especially in leucæmia), and in various other places, and which, according to Schreiner, consist of the phosphate of an organized base.

(*d*) *Schizomycetes. Tubercle-bacilli. Micrococci of Pneumonia.*—Schizomycetes appear in large numbers in the sputum because of its mixture with the oral fluids; in addition to these there are forms that come from the respiratory organs themselves, as in putrid and diphtheritic bronchitis, in purulent catarrh of the bronchi and trachea, etc. We are, however, not yet in a condition to find our way amid the crowd of different forms; the supposed "discoveries" of the fungus of whooping-cough and measles, the micrococci of diphtheria, etc., have never possessed the slightest significance to experts in this department. On the other hand, the discovery of tubercle-bacilli in the sputum by Koch is of the highest importance. For it has been established by the use of Ehrlich's method of staining that *specific bacilli are rarely absent from the sputum of phthisical patients, and that, on the other hand, their presence furnishes an absolutely safe criterion of tubercular affections.* The staining process, which at once shows the bacilli and proves their specific character, is the same that was previously described for tubercle-bacilli in sections. The dried and warmed preparation of sputum is stained for twenty-four hours on the cover-glass with a concentrated solution of gentian-violet or fuchsin in aniline-water (that is, a saturated, filtered solution of aniline); if the staining is carried on at a high temperature only a short time is necessary. Decolorization is then effected by means of strong mineral acids, as twenty-per-cent. hydrochloric, or better, in alcohol to which three per cent. of nitric acid is added. All the other schizomycetes of the sputum which were originally stained at the same time are then decolorized, as well as the masses of mucus, etc., *but the tubercle-bacilli alone appear as deeply-stained rods.* We generally meet with them in the sputum in a condition of active spore-formation; the spores appear as bright globules which include the entire breadth of the bacillus. If there are several spores in a single bacillus, the latter may on this account appear to be broken up into a row of granules. In order to bring out the contrast better we can subsequently, by any desired method of staining, add a ground color, that must naturally possess a shade which offers the greatest possible contrast to the stained tubercle-bacilli; I find that this double staining is unnecessary in most cases.

THE IMPORTANCE OF THE DISCOVERY OF TUBERCLE-BACILLI. *Severe and Light Cases of Pulmonary Phthisis.*—It is desirable to add a few observations concerning the diagnostic and prognostic importance of the discovery of tubercle-bacilli in the sputum, in which a short excursion into the field of the general pathology of phthisis can not be avoided.

Pulmonary phthisis belongs to the province of tuberculosis. The separation of cheesy pneumonia and bronchitis from the real tubercular form of phthisis, as insisted upon by Reinhardt and Virchow, must naturally be regarded as correct from the stand-point of descriptive anatomy, but the radical division of these processes, simply on the ground of their histological relations, must be abandoned, especially from an etiological and practical point of view. More accurate histological examinations showed that caseous pneumonia and bronchitis possessed the same elementary structure as so-called genuine tubercles. The author demonstrated this positively in the year

1873, in a lecture on "Local Tuberculosis" (Volkmann's "Sammlung klin. Vorträge"). *In every case of phthisis, even in the forms that are apparently nontuberculous, we always find the characteristic submiliary, non-vascular nodules (containing giant cells), which represent the chief type of tubercle after it has attained its full development.* These facts, hitherto unknown, of course limited very essentially Virchow's doctrine concerning the duality of phthisis, which had already encountered serious difficulties; the occurrence of a tuberculous pleurisy, for example, frequently observed during the course of a cheesy pneumonia, must thenceforth be regarded no longer as the invasion of a new disease (as Virchow and his school taught), but as little more than the extension to the pleura of the same pathological process which had already taken place in the lungs.

The simplicity and clearness of this idea as compared with the earlier doctrine are at once apparent, yet it was a long time before it obtained recognition. Several years later Charcot and his school, reasoning from the same observations, defended the "*unité de la phthisie*" in a series of works; Rindfleisch took a similar position. Through the later experiments of Tappeiner, Cohnheim, Salomonsen, and others, which were founded upon Villemin's discovery of the inoculability of tuberculosis, and especially through Koch's remarkable discoveries, it was proved with the greatest certainty that phthisis and tuberculosis depend upon one and the same etiological factor; at the same time it is not denied that, under certain conditions, other elements may play a part in them.

*Pulmonary phthisis is in fact to be regarded in the majority of cases as a local tuberculosis of the lungs. If, then, we find in the sputum that parasite which we know to be the cause of tuberculosis, the positive conclusion to be drawn from this is that a tubercular process is present in the respiratory tract, including the mucous membrane of the mouth and throat.*

Shall we therefore draw the further conclusion that the patient in question has developed general tuberculosis, and is in consequence beyond hope? No, that would be a gross error. The bacillus of tuberculosis is not only found in those instances in which the disease advances more or less rapidly, and eventually spreads over the whole system by extension to the blood and lymph-vessels, but also in cases which remain localized for a long time (for years, or even for decades) and finally may become completely healed, as well as in cases of "local tuberculosis." Man differs essentially in this respect from rabbits and Guinea-pigs, which are principally used in experiments; if a tuberculous affection occurs anywhere in these animals, in the eye for example, after inoculation of the anterior chamber, the tuberculosis seems without exception to spread throughout the various organs in the course of a few weeks or months; as a rule the animals rapidly succumb to the disease. Dogs appear to act differently; but as yet few facts have been presented on this subject. On the contrary, it is perfectly certain that in man the affection caused by the tubercle-bacilli may in many cases remain for several years comparatively benign and localized, or may become more or less completely cured; however, as long as the process continues, the danger is always present that it may suddenly increase in intensity and, without our being able to demonstrate the causative agents, may extend rapidly, either in a local or general manner. The human organism seems to furnish in most instances only a moderately favorable field for tubercle-bacilli, so that they generally multiply very sparingly; if under certain conditions, which are still but little known, a rapid development of the parasites takes place, this causes a speedy advance in the process. Unfortunately we are not able at present to control those conditions through which the increase of the microbe is frequently retarded or prevented; if we could attain this end, we should possess at once the therapeutics of tuberculosis. We should cherish the hope that even this lofty goal is not unattainable. At any rate, the paths which lead to it have been rendered smooth. If, too, we find tubercle-bacilli in the sputum, we infer from these the presence of a tuberculous process which, by reason of extensive, or rapid local, disturbances and the invasion of other organs, may be extremely dangerous; but the possibility of a very slow, harmless course, and even of healing, is likewise present.

By the discovery of tubercle-bacilli the diagnosis of a tuberculous process in the lungs is now rendered possible in many cases in which it was previously impracticable or very difficult. The bacilli occur in great numbers upon the walls of even the most minute phthisical cavities, in tuberculous ulcers of the bronchi, etc., and on account of their remarkably sharp outlines, through their capacity for isolated staining, they are found much more easily and quickly in the sputum than the elastic fibers, the presence of which formerly constituted the only proof of a destructive process in the lungs. By careful examinations of sputa, therefore, we shall recognize as phthisis those extremely numerous light and favorable cases which were formerly regarded as "suspicious" pulmonary catarrhs, bronchitis, etc., just such cases, in fact, as cause insignificant subjective troubles. And how very frequent such favorable cases of "phthisis" are, we learn best at the post-mortem table. In nearly one-half of the cases of healthy, powerful adults, who have died as the result of accident or acute disease, there are found on careful inspection traces of destructive phthisical processes in the lungs, in the form of caseous masses, often incrusted with lime, and cavities surrounded by firm, slaty and cicatricial tissue. Many of these cases have been entirely, or almost entirely, latent throughout their course; at any rate, in the majority of them the suspicion of a serious pulmonary affection has never been aroused; in every instance, however, tubercle-bacilli could at some time have been demonstrated in the sputum. It is well known to every practitioner, and has been frequently confirmed by clinical and anatomical facts, that a tuberculous pleuritis, or even a fatal tuberculous meningitis, may suddenly develop from phthisical centers which either appear to have healed perfectly, or have been entirely latent, and which need not be very extensive. *Whenever tubercle-bacilli are found in the sputum, the prognosis given should be a serious, but not necessarily a bad one, unless there is additional evidence.* It is certainly known that extensive phthisical processes in the lungs are arrested under favorable circumstances, and that every incipient phthisis need not go on to destruction of the organs. However, the diagnosis of tuberculous disease will always have a positive influence upon the regimen of the patient. It is highly probable that, through the early recognition of tuberculosis, the lives of many patients can be preserved, by placing them under such climatic conditions as have been proved by experience to exercise a favorable influence upon incipient phthisis.

Further investigations alone must teach how far we can decide concerning the extent of the phthisical process from the number of bacilli that occur in the sputum.

*On the other hand, the absence of tubercle-bacilli in the sputum, if this is constant, is to be regarded as a sure sign that destructive tuberculous processes are not present in the lungs at the time.* If elastic fibers occur in the sputum, while at the same time the tubercle-bacilli are absent, some other degenerative process must be inferred, such as the formation of an abscess, the breaking down of a tumor, etc. It should also be observed that there are certain chronic ulcerative conditions of the lungs which are not of a tuberculous character, and in which bacilli are not found; these cases are extremely rare. Such observations have been made by Riegel in diabetic lesions of the lungs, but, as a rule numerous bacilli are present in diabetic phthisis. The anatomical appearances of lungs from which the parasites are absent may resemble closely those of ordinary tuberculous phthisis.

It is self-evident that one should approach the examination of the sputum for tubercle-bacilli only when he is provided with the best aids. In favorable preparations, however, they may be recognized even with rather low powers; and if they are numerous, they may sometimes be distinguished at once with the naked eye through the difference in staining. But it would be entirely a mistake to attempt to search for tubercle-bacilli with any except the best immersion-lenses, for it is quite possible, according to the author's experience, that bacilli which are present in the specimen and were not found with weak systems (dry lenses, for example), were overlooked, and appear clearly only under correspondingly higher powers—at least six hundred. It is never possible to give a negative opinion with reference to tubercle-bacilli without the use of strong, excellent immersion-lenses (preferably oil-immersions), and of course Abbé's apparatus. He who dreads the expense necessary in providing these rather expensive aids, or who shrinks from the trouble which is required in working with them, must renounce the idea of examining schizomycetes.

*Micrococci of Pnemonia.*—It is known that a peculiar structure, or capsule, can be demonstrated in the micrococci of acute croupous pneumonia, which comes out with especial clearness on staining the dry preparation with gentian-violet or fuchsin. The capsule is in this way colored less deeply than the coccus itself, and usually presents a sharp outline at its outer edge. It is still doubtful if this peculiar formation can be made use of for purposes of diagnosis; several observers have already obtained results in this direction which seem to be positive. But encapsulated micrococci are sometimes seen in the sputum when no pneumonia is present, so that caution is advised. There frequently appear around micrococci colorless spaces, which have very little significance, while the pneumococci possess capsules that may be stained.

3. Pus.—Pus consists in general of a fluid (serum) in which are suspended small round cells, the pus-corpuscles; it also contains micro-organisms in most cases.

(a) *Pus-Corpuscles and Fatty Degenerated Cells.* Pus-corpuscles are very similar to, or identical with, the white corpuscles of blood and lymph. If examined when fresh, that is, in pus which has only been evacuated for a short time, they show amœboid movements, and then possess the characteristic glistening appearance of living protoplasm. In the majority of instances, however, they are already dead and their protoplasm has then become coagulated in the form of coarse granules; the nucleus (or nuclei) is usually concealed. Many very fine fat-granules are seen scattered throughout the protoplasm, and these are abundant in such pus-corpuscles as have been dead for some time, especially in the contents of so-called cold abscesses.

Pus-corpuscles are generally of nearly equal size; that is, about as large as medium-sized white blood-cells. But larger cells, which usually have vesicular nuclei, are frequently mingled with them; if numbers of fat-granules are collected in these, the well-known fat-granule cells are formed. The latter, also, are occasionally alive and perform amœboid movements. They are recognized with a weak lens as rather large, dark masses between the ordinary pus-cells; their dark outline, when viewed with low powers, induces the beginner not infrequently to think of pigmentation, although it is entirely due to the many superposed fat-granules.

The light which comes from below is, by reason of its many transitions from the fluid into the fat-drops, and from the latter back to the fluid or protoplasmic substance, reflected as if from the surface of spherical mirrors, so that it does not reach the eye of the beholder; hence the impression of black.

The same fatty cell, if illuminated from above, appears of a clear white color, likewise in consequence of its numerous reflecting surfaces; if, then, the light coming from the mirror is shut off, the fat-cells look like glistening white spheres in the dark field, it being assumed that the light from above can generally impinge upon the preparation. Strong objectives must usually be approached so close to the specimen that it is completely shaded. However, there are also real pigment-granules, which are mostly inclosed in cells, often in pus, as the remains of blood-extravasations; they are distinguished by their brown color. Hæmatoidin-crystals are sometimes found also.

(*b*) *Foreign substances.*—Pus originates either

from a free surface, as a mucous or serous membrane, an ulcer, etc., or from the midst of the tissue; in either case it often has mixed with it elements from its place of origin. Since the latter is frequently entirely unknown, and is the object of search, it is evident that the foreign substances found in pus possess at once great diagnostic value.

The important question whether an abscess is connected with bone or not is often decided by a careful microscopical examination; the irregular fragments of bone which are found in pus in these cases, and which often possess indented surfaces of absorption—the so-called Howship's lacunæ—then furnish absolutely certain proofs. They are very sharply defined through the marked glistening appearance of their calcified basement substance, and through their characteristic stellate, or spindle-shaped, bone-corpuscles; they may often be found with a low power. If necessary the preparation is cleared up with a solution of potash, so that the pus-cells disappear; if the pus is not too thick it should be allowed to settle, and the sediment should be examined.

In other cases we may find bits of food, whereupon we infer the presence of a communication with the alimentary canal; or epithelium, tumor-elements, etc., are encountered, so that we are often enlightened to a surprising extent in this way with regard to the diagnosis, while the information is of direct value from a therapeutical stand-point. As an example: In the pus contained in an abscess which developed beside the thyroid gland of a woman, after confinement, there were found large numbers of flat epidermal cells and a good deal of cholesterin; the diagnosis of suppurating bronchial cyst was at once made, and the sac was extirpated. Suppurating echinococcus-cysts, also, are frequently recognized correctly only by a microscopical examination of the pus, in which either entire scolices, the characteristic hooks, or the stratified, homogeneous membranes are demonstrated.

(c) *Schizomycetes and Actinomycetes.*—Micrococci are found in very large numbers in pus from acute abscesses,* being generally arranged in the form of chains, which are visible either when examined in the fresh state without further treatment or after they have been stained by the familiar processes. Microorganisms do not constantly occur in chronic suppurations; reference is, of course, made only to those cases in which the pus does not communicate with the external air, since saprophytic organisms are then always present.

In tuberculous abscesses (periarticular and articular abscesses, fungous arthritis, strumous and carious suppuration, cheesy and purulent lymphadenitis, etc.) tubercle-bacilli are naturally often observed, and they always possess a pathognomonic significance under such circumstances.

On the other hand, they are not so invariably present as in pulmonary phthisis. Bacilli can not be found in many genuine tuberculous abscesses; according to Schlegtendal ("Fortschr. d. Med.," 1883, S. 537), they are met with in only about one-half of the cases.

Less often other organisms appear in pus; from

* Comp. Ogston, "Über Abscesse," "Arch. für clin. Chir." Bd. 25.

these are excluded those which are purely accidental and develop subsequently—for example, the micrococci which cause the color of "blue pus."

The actinomycetes—which were found in man, first by Langenbeck, and later by J. Israel, Ponfick, and others, and in cattle by Bollinger*—may be recognized microscopically in pus as gelatinous, miliary granules; on squeezing the granules their peculiar structure at once becomes apparent, so that staining is unnecessary.

4. THE URINE.—In discussing the microscopical examination of the formed elements of the urine we consider first

(a) *Precipitates and Crystals.*—The urinary salts, especially urate of sodium, which are normally held in solution at the temperature of the body, are precipitated (if they are present in considerable quantities) as soon as the urine becomes cool, in the form of fine, often rather irregular, granules. The author has frequently remarked that these are regarded as micrococci by novices, their quivering molecular movement being mistaken for vital locomotion.

The salts of uric acid readily dissolve on warming slightly, also on the addition of acids, through the action of which uric acid separates in the form of characteristic crystals, generally prismatic in shape, and frequently of a brownish color. In febrile conditions, as in gout, urate of sodium is usually much increased; the uric acid often separates spontaneously in the urine a short time after it is voided, that is, when no acids have been added. This was formerly

* Comp. Ponfick, "Die Actinomycose," Berlin, 1882.

described incorrectly as "acid fermentation of the urine." The process is simply this: The acid phosphate of sodium, under the decomposing action of the urate, is changed into basic phosphate of sodium, while the uric acid is set free. A genuine acid fermentation of the urine, with an increase in its acid reaction, occurs in diabetes (Voit and Hofmann).

Simultaneously with the appearance of the uric acid crystals there often occurs a separation of oxalate of lime in the form of small, shining octahedra, which, when viewed from above, resemble envelopes. When the urine is abnormally rich in oxalates, we speak of "oxaluria." Oxalate of lime, as is well known, forms the principal component of an important class of urinary calculi. In the urine after excretion from the body there occurs regularly an alkaline, or ammoniacal, fermentation, that is a decomposition of urea to form carbonate of ammonium, which change is effected through the agency of an amorphous ferment which was isolated by Musculus. This ferment, however, is itself always produced by micro-organisms, that is, by schizomycetes. If these organisms, or their germs, get into the urine that is contained within the bladder (this they do principally during catheterization) the alkaline fermentation may take place at once in the bladder, especially if the urine is retained, as in vesical paresis. During alkaline fermentation the urine becomes very turbid. This turbidity (aside from the presence of schizomycetes) is due to

1. Phosphate of ammonium and magnesium (triple

phosphates), in the well-known form of coffin-lid crystals, which are immediately soluble in acids.

2. Urate of ammonium, in the shape of "morning-stars," or balls covered with fine projections.

3. Phosphate of lime, an amorphous precipitate. In certain pathological conditions, beside the substances already mentioned, which may be present in enormous quantities ("urinary gravel"), various other crystalline and granular precipitates are found, such as regular hexagonal tables of cystin (in cystinuria), xanthin and allied bodies, tyrosin in the shape of needles gathered into sheaves, and generally of a yellowish color (especially in acute yellow atropy), sulphate and carbonate of lime, etc.; all these precipitates can generally be very easily distinguished by means of simple micro-chemical reactions. Chemical text-books must be consulted for information upon this subject.

(b) *Urinary Casts.*[*]—Three principal varieties of casts are to be distinguished, hyaline, waxy, and brown. The hyaline consist of a substance which has a very delicate outline and is perfectly transparent, so that they may be easily overlooked; they are sometimes rendered more apparent by reason of adherent fat-granules. Their breadth frequently only equals the diameter of a red blood-corpuscle, but as a rule it exceeds it. They are found in albuminous urine during the most various pathological processes, even in cases in which neither inflammatory, nor other

---

[*] Casts were discovered in the urine by Henle, and, as a result of examinations made on the cadaver, they were soon recognized as molds of the urinary tubules. "Zeitschr. für rat. Med.," Bd. 1.

similar conditions are present in the kidneys, as in many febrile affections, in icterus, etc. They are accordingly to be regarded as the attendants of every case of albuminuria, though mild in character. The waxy casts, on the other hand, when found in considerable numbers, are of positive diagnostic importance; they should always be regarded as sure signs of disease of the kidneys, and may accompany congestion as well as genuine nephritis.

Waxy or colloid casts are composed of a substance which presents sharp outlines and is more or less glistening, or they may be slightly clouded by deposits of very fine granules in their interior; in the latter case they are distinguished as a special variety, receiving the name "granular" casts.

They are of quite variable width, reaching five one-hundredths of a millimetre and more; their shape is, as a rule, perfectly cylindrical, or cruciform on transverse section, while they have sometimes irregular, serrated borders, especially in acute nephritis. They are often filled with small round cells and fat-drops, and sometimes also with epithelial cells; there are forms of casts, in fact, which consist almost entirely of epithelial cells that are more or less firmly coherent (epithelial casts). The old expression "fibrinous" cast has properly been given up by all; the substance composing these casts is essentially different from fibrin, since they are neither redissolved by acetic acid, nor do they swell when placed in it, but they merely tend to lose their glistening character and dark contours, and finally their granular contents, through the action of the acid. They then appear as pale hyaline casts. Casts are stained a slight yellow color by the action of iodine, waxy casts frequently becoming even dark yellow or reddish brown.

We shall not enter here into the controversies regarding the origin of casts; the hyaline appear to be a direct product of exudation, while the waxy may, at least in part, come from disintegrated epithelial cells.

A peculiar form of small, brownish casts was found in the urine by Riegel* during the first few days after the occurrence of fractures; he explains these, with great probability, as products of the fibrin-ferment which has got into the circulation. In these cases there are usually found also larger or smaller masses of fat, which collect in the form of small drops in the upper layers of the urine. This fat is introduced into the circulation through the agency of the wound, forms emboli in the renal vessels, and is then gradually excreted with the urine. In renal hæmorrhages and hæmorrhagic nephritis there are often observed brown casts, which are stained with blood coloring-matter, and frequently also real blood-casts.

(c) *Pus- and Mucus-cells. Epithelial Cells.*—Lymphoid cells, pus- and mucus-corpuscles often appear in the urine; they may come from the kidneys or urinary passages, or they may originate in an abscess which has ruptured somewhere into the urinary tract. Epithelial cells also are often found, the source of which can not always be positively determined. In urethral and vesical catarrh, epithelial cells are sometimes

* "Über das Verhalten des Urins bei Knochenbrüchen." "Deutsch. Zeitschr. für Chir.," Bd. 10.

found, within which one or more lymphoid cells are inclosed; it was formerly supposed that these arose in an endogenous manner, but they are now regarded as having been invaginated, that is, they have penetrated subsequently into the interior of the epithelial cells (Volkmann, Steudener). Cells containing fat-granules appear but seldom in the urine; Leyden found them in acute nephritis.

(*d*) *Tumor-elements* are not difficult to recognize if some care be used, but this only applies to the observer who has become thoroughly acquainted with the epithelial elements of the kidney and urinary passages. The extremely multiform epithelial cells of the bladder (sometimes of very large size), which are provided with several nuclei, have often been mistaken for cancer-cells.

Diphtheritic and tuberculo-caseous masses appear in the urine; they generally come from the bladder.

(*e*) *Entozoa* occur with extreme rarity in the urine; these are echinococci, as well as *filaria* (the latter have hitherto been found only in the tropics in cases of chyluria), also the ova of *Distoma hæmatobium*. Many errors in observation have occurred in this field; a certain author, for example, described the ova of *Strongylus gigas*, which he professed to have seen in the urine, but on closer examination it turned out that he had mistaken for ova the granules of *Semen lycopodii*, which had got into the specimen through want of cleanliness.

(*f*) *Vegetable Parasites.*—Sarcinæ are found in the urine only in very exceptional cases.

## THE EXAMINATION OF FLUIDS. 165

In order to prove the presence of bacteria and micrococci in the urine, it is of course necessary to examine a perfectly fresh specimen; normal urine is always free from such organisms, but within a few hours after it has been voided they are found in great numbers. They are observed in the largest quantities, as was first proved by Traube, in freshly passed urine, in those cases in which alkaline decomposition has occurred in the bladder after catheterization, and in consequence cystitis has developed. There are then found multitudes of triple phosphate crystals, lymphoid cells, and immense swarms of rods and granules which stain deeply with basic aniline dyes, and are often rolled together in large zoöglœa-masses. In these cases the schizomycetes, the germs of which have entered the bladder by means of the catheter, very often wander through the ureters into the renal pelvis and true parenchyma of the kidney, and are then found in the interior of pyelo-nephritic abscesses (Klebs). In different infectious diseases, especially in metastatic abscesses, the organisms make their way from the blood into the urine, as can be demonstrated with certainty; as yet, however, few reliable clinical investigations on this subject have been reported.

Those forms of schizomycetes which furnish the ferment that decomposes urea, never, as far as is now known, pass from the blood into the renal excretion (probably because they do not exist in the blood), but are always introduced from without. On the other hand, they must be present in the alimentary canal, for the urea which has entered the intestine (in cases of uræmia) is rapidly transformed into carbonate of ammonium.

The gonococci, which occur in the urine in gonorrhœal cystitis, do not decompose the urea.

I have several times demonstrated the presence of tubercle-bacilli in the urine in tuberculosis of the kidneys and urinary tract, though at first only in the cadaver; later they were frequently observed and utilized for purposes of diagnosis. The prognosis of tuberculous affections of the urinary organs is generally a very unfavorable one.

5. SECRETIONS OF THE GENITAL TRACT. (*a*) *Vaginal Secretion.*—The secretion of the vagina is a fluid which contains more or less abundant large, partially corneous, epithelial cells, and also some round cells. The latter vary in dimensions from the size of white blood-corpuscles up to forms which are four or five times larger; the larger round cells generally contain numbers of fat-granules.

A great many micro-organisms appear in the vaginal secretion; all the conditions for the rich development of both harmless and injurious parasites are present here, just as in the mouth. Among the harmless ones we must always include the *Trichomonas vaginalis*, discovered by Donné in the vaginal mucus, an infusorium provided with flagella and cilia and capable of active motion. Mold-fungi also develop upon the vaginal mucous membrane, especially in pregnant women; if they form in larger masses they cause white patches and a slight inflammation of the membrane. This, according to Haussmann, is the *Oidium albicans*, or thrush-fungus, through infection with which the thrush of new-born infants may be produced. Many of the schizomycetes that appear in the vaginal secretion have hitherto not been distinguished from one another at all, or only in an imperfect way. With reference to the micrococci of gonorrhœa, see below under *c;* these, too, have not yet been so exactly described that they can aid the diagnosis in doubtful cases.

(*b*) *Fluids from the Uterus. Dysmenorrhœal Membranes. Decidual Remains. The Diagnosis of Carcinoma of the Uterus.*—Beside the normal mucous plug of the cervix uteri, which contains only a few lymphoid cells, there is found during inflammatory conditions a fluid secretion of the uterine mucous membrane, often purulent, in which, in addition to lymphoid cells, many cylindrical epithelia appear, generally without cilia.

The menstrual discharge consists mostly of blood; the lochia also contain numerous elements from the placental remains, especially from the deepest layer of decidua, which are retained in the uterus after delivery. The epithelial cells from the lower ends of the glands, with their bright, almost vacuolated nuclei, are especially characteristic.

If portions of the placenta have been retained we frequently find in the lochia smaller or larger pieces of it; the structure and form of the dendritic, branched placental villi (chorionic villi) are so characteristic, that they are always recognized as such at once. Deposits of lime are frequently observed in such placental remains, so that occasionally even firm concretions are expelled from the uterine cavity; a histological examination of these at once furnishes a clew to their origin.

In certain forms of dysmenorrhœa, as is well known, membranes are cast off, while at the same time the patient often has pains resembling those of labor. Examination of these dysmenorrhœal membranes invariably shows that they consist of actual pieces of the uterine mucous membrane; even the sacciform glands and their openings upon the lining surface of the uterus can be demonstrated. Such membranes were formerly called "menstrual decidua," and the question was also raised whether in these cases there was not really an early abortion. The histological examination decides if this is the fact or not. For the minute structure of the decidua of pregnancy, that is of the uterine mucous membrane as changed by conception, is clearly different and is perfectly characteristic of the condition. . The interglandular tissue of the mucous membrane of the uterus always consists, as far as is at present known, of small round cells as large as lymphoid cells, containing very little protoplasm, and this is the case under all circumstances, as in the swollen state of menstruation, in dysmenorrhœal membranes, in the different varieties of endometritis, in the swelling due to uterine myomata, etc. Pregnancy alone causes a characteristic change in the cells; even at the beginning of this condition we find in the swollen mucous membrane the familiar large decidual cells, from five to ten times the size of leucocytes, rich in protoplasm, round or polygonal in form, and provided with processes. The cells of the decidua retain this size and shape until the end of pregnancy. The decidual tissue, if the glands be excepted, resembles many forms of large-celled sarcoma. In extra-uterine fœtation, also, a swelling of the uterine mucous membrane regularly occurs, and fragments of the same are often expelled; the characteristic formation of the decidua (its large cells) can always be demonstrated in these cases.*

CARCINOMA, EROSION, OR ADENOMA?—In carcinoma of the uterus there are frequently found, suspended in the fluid which has exuded from the cancerous ulcer, cellular elements, or even pretty large fragments and shreds, the structure of which as viewed through the microscope assists in establishing the diagnosis. In doubtful cases, however, where the question is whether we have to do with carcinoma or with a simple erosion, the examination of the secretion alone will never be sufficient; in such instances small bits are frequently excised, by the histological study of which the diagnosis should be confirmed. We shall introduce here a few observations upon this subject because of its practical importance.

The floor of an erosion consists of granulation-tissue, which is usually covered by several layers of epithelial cells; gland-like prolongations provided with lumina extend from this epithelium into the depths of the granulation-tissue; these may, however, be solid epithelial columns, which subdivide and unite with one another to form an irregular network.

As is at once evident, this structure closely re-

* Wyder, "Arch. für Gynäkologie," Bd. 13.

sembles cancer, although it occurs in perfectly simple benign erosions. A careless observer who, without further evidence, makes the diagnosis of cancer, may readily occasion in this way a great deal of unhappiness; he will undertake mutilating and dangerous operations in cases in which a radical extirpation is not indicated. Another element must be present before we are justified in making the momentous diagnosis of cancer. The secondary epithelial proliferation, which extends into the granulation-tissue and is often perfectly atypical, is not limited to erosions of the uterus, but appears very often and in the most dissimilar localities, as in the skin, liver, lungs, etc.; it may occur at any spot where granulation-tissue comes into direct contact with epithelial surfaces. The atypical epithelial outgrowth does not vary at all with the character of the original affection which has led to the formation of the granulation-tissue; the process is a perfectly benign and harmless one, and would be of but little interest to us from a practical stand-point, if the structure of cancer, especially at the beginning of its course, did not bear the most perfect resemblance to the innocent atypical growth of epithelial cells.

Cancer has even been defined, from a histological point of view, as "an atypical proliferation of epithelium" (Waldeyer). This definition, however, is not sufficient; we must add to it, as being necessarily the most important peculiarity of cancer, the words "of a malignant character." With this expression we leave the purely histological field, for we can not recognize directly the "malignant character," either in the cell or in the tissue. Nevertheless the microscopical examination again proves the fact of malignancy; *for the malignancy of the process is shown if it forces its destructive way through various tissues without hindrance*, while an innocent growth remains limited to the tissue from which it originated, and either leaves the neighboring parts entirely intact, or merely pushes them aside. If we discover in the uterus, for example, that the process is not confined to the mucous membrane, but that it also invades the muscular tissue, and that the muscle is partly replaced by granulation-tissue,* traversed by processes of atypical epithelium, we have before us a clearly malignant element, and then only do we make the positive diagnosis of cancer.

Fragments removed for the purpose of histological examination must also include at least a part of the muscular layer; the diagnosis of cancer can not be surely established unless it has been shown that the muscle is affected. In this respect I am opposed to C. Ruge, † who proposes a very simple method of distinguishing a cancerous ulceration from an innocent erosion.

Ruge affirms that in cancer the epithelial columns are solid, while in simple erosions they contain lumina, and this is certainly correct in many instances. But if he means that this difference is present without exception, and that use can therefore be made of it for purposes of diagnosis, he is entirely in error; for, on

---

\* The stroma in young cancerous growths is usually composed of granulation-tissue.

† C. Ruge, " Berlin klin. Wochenschrift," 1878.

the one hand, in many cases of genuine malignant cancer the cell-columns possess most beautiful and regular lumina, and, on the other, solid epithelial plugs are very often observed in innocent erosions, as may be readily shown by examinations of bodies. The criterion proposed by Ruge must accordingly be regarded as entirely unreliable for the diagnosis of cancer.

These maxims with regard to the decision of this question should be observed in the case of all the other organs; as long as we find atypical epithelial growths in localities in which epithelium has always existed, as in the skin, mucous membranes, in all true glands, etc., we must always show positive proof of malignancy before we make the diagnosis of cancer. If, however, these atypical epithelial processes occur elsewhere—in muscle, for example, in bone, or in the lymph-glands—then the eroding, or metastatic, character of the affection has been demonstrated at once *eo ipse*, and the diagnosis of cancer is necessarily made.

. Many authors use the term epithelioma as synonymous with cancer, which, according to my opinion, is not to be commended. The word "epithelioma" we should retain as the general designation for *epithelial tumors of all kinds;* we have the word "cancer" for such of them as are malignant, while the expression "adenoma" is best applied only to benign growths.\*

The cause of the malignancy of cancer is not yet known. Nevertheless, as the results of thousands of

---

\* Epithelial elements are found in teratoid tumors, even in the midst of other tissues, without the growths being considered as malignant. In myomata of the uterus cysts have rarely been discovered which were lined by ciliated epithelium.

observations, we can at once, and with great certainty, give a fatal prognosis in all cases in which it was necessary to make the diagnosis of cancer according to the principles before stated; if total extirpation is not performed the patient will soon die.

To this practical rule there are only very rare exceptions; there are certain forms of superficial cancer of the skin (rodent ulcer) which may run an extremely protracted course.

On the other hand, we can deny the presence of *immediate* danger if the tests of malignancy already detailed are absent; it is, of course, not impossible for a cancer to develop subsequently from a growth that was originally innocent, but positive danger is only imminent when the malignant character of the process has been demonstrated.

The diagnosis may, under certain circumstances, become difficult, as when an atypical epithelial growth is combined with a syphilitic or tuberculous ulceration; special caution is advisable in such cases, and it is better to first await the result of vigorous local treatment.

(c) *Gonorrhœal Secretion.*—Neisser[*] found a specific variety of micrococcus in gonorrhœal pus, which is distinguished by its forming small groups, in which the single granules are relatively far apart; they frequently lie upon the exterior and in the protoplasm of pus-corpuscles. The same micrococci are found in the secretion of specific conjunctivitis. They are readily stained in the usual manner with aniline dyes. The characteristic points about the gonococci are not

[*] Neisser, " Med. Centralbl.," 1879.

yet sufficiently clear, so that we are not in a position to distinguish with certainty the micrococci of gonorrhœa when they are mingled with other microorganisms, as in the vaginal secretions.

(*d*) *Semen and Prostatic Secretion.*—The highly characteristic, active spermatozoa are found in the semen; they may still be demonstrated very well, as a rule, in dried semen (semen-stains) by moistening it with salt-solution. If we find in doubtful cases small glistening bodies which look like the heads of spermatozoa, and also fine threads resembling tails, but no perfect spermatozoa, we must not be led into the error of drawing a positive conclusion from the fragments discovered. Similar small corpuscles, or threads, are easily shown in dry spots of any character whatsoever, and they may have every possible source; the nature of the semen-stain is only positively recognized when perfect seminal bodies, with heads and tails *in continuo*, can be demonstrated. If spermatozoa are not found in the semen, care should be taken to distinguish the temporary from the permanent form of azoöspermia (*sterilitas masculina*); if several ejaculations are provoked, one soon after another, the fluid, according to the testimony of a number of observers, is finally entirely devoid of spermatozoa. The secretion of the testicles is then temporarily exhausted and the fluid ejaculated comes only from the seminal vesicles and the prostate.

The prostatic secretion is generally mingled with the semen, but it is sometimes discharged independently by pressure upon the prostate, as during defecation. It often contains, beside the laminated amyloid corpuscles, a large number of octahedral or lancet-shaped crystals, which, like the asthma-crystals in bronchial secretions, represent the phosphatic salt of an organic base—the so-called Schreiner's base. They have long been known in the semen as spermatic crystals; Fürbringer * has shown that they originate in the mixed prostatic secretion and that they are the source of the peculiar odor of the semen. They may often be demonstrated in freshly discharged seminal or prostatic secretion, but otherwise only after it had stood for some time; they always form in large quantities if ammonium phosphate is added to the seminal or prostatic fluid.

6. CONTENTS OF THE STOMACH AND INTESTINE.—
The microscopical examination of vomited matters and of the stools has long been practiced, and in many cases it furnishes important diagnostic aids.

(a) *Remains of Food* are, of course, always found in large quantities. During a meat diet striated muscular fibers occur regularly in the dejections (Frerichs); likewise the cellulose membranes are found during a vegetable diet, while, under the normal conditions of digestion, starch is either not present in the fæces, or only in very small amount. Most of the starch-granules are at once perfectly extracted in the stomach; they then no longer stain blue with iodine, but take only a light-yellow color. Whenever a considerable quantity of starch is found in the evacuations of the intestine, a pathological condition is present, generally attended with diarrhœa.

Among the animal food-elements tendons, apo-

* Fürbringer, " Zeitschr. für klin. Med.," Bd. 8.

neuroses, fragments of large arteries, etc., long resist the digestive forces; large quantities of such materials are often found, not only in the dejections of insane, or especially gluttonous individuals, but in the stools of men who are quite prudent in their eating. These masses, occurring in the form of so-called "intestinal concretions," not infrequently occasion serious anxiety both to the patient and to the physician; but on microscopical examination it is at once recognized that this is unfounded. Undigested vegetable matters are often voided; Virchow has called attention to the frequent appearance of orange-pulp in the stools.

(*b*) *Epithelial Cells, Mucus, etc.*—The epithelial and gland-cells of the alimentary canal are frequently met with in the contents of the stomach and intestine (but generally much altered), without this discovery possessing any particular significance. Large masses of mucus, which are apt to contain leucocytes, mucus-corpuscles, etc., denote the presence of a gastric or intestinal catarrh.

Sometimes, in addition to the mucus, collections of fibrin appear in the stools, either in the form of membranes or of dendritic, branched processes that unite in the shape of a net-work. These are to be regarded as the products of a pseudo-membranous inflammation of the colon which, as is well known, sometimes assume the appearance of a net-work. The evacuation of these masses, which may attain considerable dimensions, is sometimes attended with severe pains like those of labor. They usually contain fibrin, as well as mucus, and therefore only partially dissolve in acetic acid; nothing else is found in them except round cells, or their *débris*.

(*c*) *Entozoa* of different kinds appear in the alimentary canal, many of them being unimportant parasites, while others are highly dangerous. The animals themselves, or portions of the same, are found in the stools, but their ova are valuable as a means of recognizing them in the dejections; accurate descriptions and illustrations of these will be found in text-books upon pathological anatomy, also in the well-known works of Leukart, Braune, and Davaine.

In the year 1880 Perroncito made the important discovery that the "tunnel-disease," which decimated the St. Gothard workmen, was caused by the *Anchylomum duodenale* (or *Anguillula intestinalis* and *Pseudorhabditis stercoralis*); the eggs of these parasites are found in immense numbers in the stools of the patients, and are perfectly characteristic. In the so-called "mountaineers' anæmia" also, which prevails in the mines of Hungary, he found the same parasites. Perhaps a careful examination of the stools in other hitherto problematical diseases will furnish new light.

(*d*) *Vegetable Parasites.*—Sarcinæ have long been recognized in the contents of the stomach, especially in cases of dilatation; no clinical importance is attached to them. Of the other vegetable parasites, micrococci, bacilli, etc., appear in the stomach only in small numbers, but very large masses of yeast-fungi are often seen.

Yeast-fungi are likewise present in the contents of the intestines and in the dejections, but micrococci and bacilli, both large and small, are found in extremely large numbers; certain forms of rods have been described by Ferd. Cohn as *Bacillus subtilis*. Among these confused masses of micro-organisms it has thus far been impossible to distinguish any specific pathogenic forms, with the exception of the tubercle-bacilli characterized by their peculiar relation toward staining processes.

There are several forms of organisms in the contents of the intestine which turn blue on the addition of iodine; of special interest is the *Clostridium butyricum* (Prazmowsky\*), large quantities of which are found in the lower section of the ileum and in the colon (but not in the upper portions of the alimentary canal), especiallly after a vegetable diet (Nothnagel †). Their size is variable, something like that of yeast-fungi, though their form is different; they appear either in the shape of rods, often drawn out into single or double-pointed extremities, or of elliptical, spindle-, or lemon-shaped bodies. They are frequently collected together into chains or groups. They are stained blue with iodine, either entirely, or only at their centers, while their extremities (or entire periphery) take a yellowish tinge. They are probably identical with the *Bacillus amylobacter* of Van Tieghem, and represent the ferment of the butyric-acid fermentation which occurs in the contents of the intestine. Nothnagel has also described still smaller forms, which occur in the dejections and may be stained with iodine. Tubercle-bacilli are very often

---

\* Prazmowsky, "Untersuchungen über Entwickelungsgeschichte und Formentwickelung einiger Bacterien," Leipzig, 1880.

† Nothnagel, "Zeitschr. für klin. Med.," Bd. 3.

encountered in cases of tuberculous ulcers of the intestines, in the contents of the gut, and of course in the fæces. Numbers of tubercle-bacilli are regularly present on the floors of these ulcerations; they then become mingled with the intestinal contents, and may be used for making a diagnosis during the examination of the fæces.* It should be remembered that the sputum of phthisical patients may be swallowed and may also give rise to a mixture of tubercle-bacilli with the fæces; in every case a tuberculous affection of the body is certainly established by the finding of these organisms. The other bacilli which appear in the contents of the intestine are at once decolorized by treatment with acids; but certain forms of large, round micrococci are observed in the fæces, which, like tubercle-bacilli, retain their color for a long time when treated with acids. These bodies are regarded by Koch as spores which possess exceptional capacities for staining, since it is well known that the spores of most bacilli with which we are acquainted do not stain. By reason of their globular shape they can certainly never give rise to diagnostic errors.

7. EXUDATIONS. THE CONTENTS OF CYSTS.—The microscopical examination for diagnostic purposes of exuded fluids, cyst-contents, and the like, is very often undertaken at the present time, on account of the great frequency with which exploratory puncture is resorted to.

The technique of these examinations is as simple as the difficulty of estimating properly the diagnostic value of the discoveries is great; it is usually neces-

* Lichtheim und Giacomi, "Fortschr. der Med.," 1883, S. 3 and 150.

sary merely to examine the sediment according to the methods already described, or to stain the dried preparation.

Large quantities of fat occur not infrequently in exudations, and cause them to assume an opalescent or milky character. The fat is sometimes contained in the fluid, as in chyle, in the form of minute, irregular, slightly glittering granules, intimately combined with albuminous bodies, so that it is not recognized as such without further tests; it is only after adding acetic acid or alkalies that the albuminates are dissolved and the fat runs together to form large, shining drops. This so-called *hydrops chylosus* is never seen except when chyle is poured out into the abdominal or thoracic cavity, or as the result of a wound, or obstruction to the flow of the chyle in the lacteals or thoracic duct.* In other cases the fat comes from disintegrated fatty cells, and is at once recognized by its small, glistening granules (*hydrops adiposus*); this is observed in chronic inflammation, as well as in carcinoma of the serous membranes.

Serous effusions also occur, which only show an opalescent or milky cloud, due to the presence of albuminous granules. Red blood-corpuscles are found in variable quantities, and often in an altered condition, as decolorized, shrunken, etc.

Lymphoid cells are almost never absent, only they are far less numerous in simple transudations than in inflammatory exudations. Active amœboid movements may be observed in them, but in many cases

* Compare Quincke, "Ueber fetthaltige Transsudate," "Deutsch. Arch. für klin. Med.," Bd. 16.

they are already dead; they are frequently filled with numerous fat-granules.

Endothelial cells appear in serous effusions partly single, partly in coherent plates, often also filled with fat or transformed into globular bodies; they are provided with one or more nuclei, and sometimes contain vacuoles also.

Epithelial cells, cylindrical, polygonal, or flat, are observed in the contents of cysts; they often furnish clews to the diagnosis of the origin of cysts. Ciliated epithelial cells are sometimes found, especially in monolocular cysts.

Tumor-elements, which are mingled with exudations, are generally quickly precipitated, and this happens even within the body, so that if punctures be made high up the cells may be missed entirely, while in the more dependent parts they are present in abundance.

This applies chiefly to cancer, the cells of which are distinguished by their variable size, their large nuclei, and their polymorphous forms; vacuoles are also seen very frequently. However, it is not usually advisable to make a diagnosis of cancer from finding a few cells; it is much better for the beginner to leave the diagnosis in doubt until the cells are found collected together into clusters or balls. Quincke* affirms that the glycogen-reaction can probably be employed in forming a diagnosis; this reaction often occurs in cancer-cells, while endothelial cells usually appear to be devoid of glycogen.

* Quincke, "Ueber Ascites," "Deutsch. Archiv. für klin. Med.," Bd. 80.

He who desires to establish such diagnoses must first examine thoroughly the formed elements of the different varieties of serous effusions as they appear in the cadaver, since otherwise errors are certain to occur. The variations of form shown by the endothelial cells and their derivatives in ordinary subacute or chronic inflammations of serous membranes are often very surprising to the novice, and may readily give rise to a false diagnosis of cancer.

Schizomycetes have as yet been seldom examined in effusions; possibly a more careful study of these will furnish other important aids to diagnosis.

In the pleurisy and pericarditis accompanying acute pneumonia, quantities of pneumococci frequently develop, which, in doubtful cases, may very materially assist in the diagnosis. By puncturing the lung-substance Günther and Leyden succeeded, in two cases of acute pneumonia, in finding micrococci in the alveolar exudation of the living subject; Günther notes especially in his case the presence of "colorless capsules" around the pneumococci. Of the animal parasites, the echinococcus is of special importance in this connection; the chitinous hooks and the stratified membranes are to be used as positive microscopical proofs. In inspecting the sediment carefully we frequently find, even on macroscopic examination, the separate or grouped scolices, appearing as white specks.

## VII.

### THE EXAMINATION OF SOLID ELEMENTS OF THE BODY, EXTIRPATED TUMORS, ETC.

THE microscopical examination of solid elements of the body, tumors, and the like, is effected according to the methods before described. The isolated elements are obtained either by examining the fluid, by scraping the cut surface, or by teasing, after previous maceration. *For this purpose always make, with a perfectly clean knife, a freshly cut surface;* otherwise you run the risk of being constantly embarrassed by numbers of accidental impurities that have collected on the surface.

Sections are also prepared from the fresh specimen by means of the curved scissors, razor, double knife, or freezing-microtome, and are examined at once in salt-solution; Bismarck-brown and solution of iodine are mostly used for staining fresh sections.

Specimens should always be examined first in a fresh condition, since this offers many advantages. Substances then show their natural transparency, and in this way we are best able to compare the histological structures which we have found with the differences observed macroscopically. If there is any fatty degeneration present it is only seen in its full extent in the fresh preparation.

The freezing-microtome should be used in all cases in which it is desirable to prepare for demonstration neat specimens that are suitable for staining. It is generally possible in this way to obtain in a very short time entirely satisfactory, perfect sections which can be examined directly in salt-solution; the nuclei are then stained very beautifully with Bismarck-brown, the specimen is transferred from Bismarck-brown to alcohol for a short time, and is then mounted in glycerin, or it is placed first in oil of cloves, and afterwards in Canada balsam. In this way, where haste is necessary, specimens affording a perfect view of the structure may be prepared within a few minutes, from organs freshly removed from the cadaver, as well as from extirpated portions of the living body, or even during an operation; this, under some circumstances, is a very important advantage.

In many instances, however, we are not successful in the examination of the fresh preparation; large and very delicate sections of fresh organs are apt to curl up, and are so extremely soft and easily destroyed that sometimes, in spite of the greatest care and much loss of time, it is not possible to transfer them intact to the slide and to spread them out well. Then the true value of hardening appears. Hardening (aside from special cases) is always effected by means of alcohol; only, in the case of the nervous system and the eye, it is sometimes necessary to give up this hardening fluid, the effects of which are simple and easily controlled, and to make use of chromic acid and its salts (Müller's fluid).

Hardening in alcohol is best accomplished by

placing small pieces of the substance in large quantities of absolute alcohol; in this way the concentrated spirit penetrates rapidly into the interior of the pieces, and, by immediately coagulating the albumins, effects a speedy fixation of the tissue-elements (compare page 24).

The hardened specimens are then always cut with the microtome (compare page 13); a large number of successive sections are thus obtained rapidly, which can be treated in any desired manner. Always examine first in simple water and glycerin, and after that study the effect of reagents and staining. In many cases a single brief examination is sufficient, since we are frequently required only to classify, or to name, a given condition; in this case it is sufficient to examine one preparation unstained, and another in which the nuclei have received any desired color. In other instances, however, we meet with unsuspected, often surprisingly new, structures and combinations; it is then advisable to store away a great many sections, preserving them in alcohol. New views, or new questions, often come up after the lapse of some time, which can be answered by some peculiar method of treatment such as had not previously been thought of.

As regards the method of procedure in isolated cases, and the value of the observations for diagnostic and pathological purposes, we should exceed the bounds of our little book if we undertook to treat this subject specially. This requires, besides clinical knowledge, an accurate training in pathological anatomy and histology, and great circumspection, which can only be acquired through a thorough experience with the subject. *It may be laid down as a fundamental principle that a discovery should never be regarded as pathological in its character unless the specimen has in every case been directly compared with the normal organ, which has been similarly treated.* The disregard of this maxim, which really sounds as if it were self-evident, has already introduced into science many gross errors, and leads practically to the most serious misconceptions. Many new facts in normal histology have been discovered in the course of pathological examinations; the endeavor was made to prove that the new objects, which were first represented to be pathological, were the cause or products of a certain disease, until it was ascertained later that these were perfectly normal structures that had been previously overlooked. This was the case, for instance, with the cells filled with fat-granules, which are found in the central nervous system of old fetuses and new-born infants; these cells were first regarded as evidences of encephalitis, until it was afterward proved by Jastrowitz and Flechsig that they are necessary transitional forms in the normal development of the white substance.

In pathological discoveries we are very often concerned with *quantitative deviations from the normal;* for example, with the proliferation of nuclei, clouding of the protoplasm (although this shows a certain amount of clouding even when normal), atrophy, and diminution in the size of cells. It is evident that in order to establish such gradual differences, a direct comparison with the normal organ, which has been treated in precisely the same way, is positively necessary. Furthermore, objects are frequently found which are not exactly normal and yet which possess scarcely any pathological value—subnormal conditions. To this class belong especially the numerous senile degenerative changes, which are almost insignificant as long as they are not present in excess. They should always be observed, without, however, too great importance being assigned to them.

The extent of the abnormal process sometimes has a very important influence upon the decision as to its pathological and clinical importance. The beginner is frequently inclined to estimate too highly the significance of his discoveries, and to regard as a very extensive change one that covers the entire field of view of the microscope. Only repeated examinations can guard the novice from drawing erroneous conclusions in this respect; we gradually learn to judge correctly concerning the extent to which a process has spread over an organ by examining different portions of the organ, and especially by a systematic use of low powers. If, for instance, several shrunken glomeruli are discovered in a kidney, do not decide at once, without further evidence, that the glomeruli are all shrunken, but first determine what per cent. of the glomeruli are altered, and how many still remain normal, and whether the affection is a diffuse one, evenly distributed throughout the entire organ, or is localized; or whether the foci are more or less numerous and the intervening parts have remained quite normal, or also appear to be somewhat changed. If only a small portion of the organ has undergone change, the process, although quite intense at the single spot in question, may be of very trifling clinical significance; it should be remembered that even the sudden extirpation of an entire kidney from an otherwise healthy organism is tolerated without any marked disturbance.

On the other hand a change that is really much less striking may, if it is diffused over the entire organ, be attended by very injurious consequences to its functions, and consequently to the whole body. This is the case, for example, in glomerulo-nephritis, a disease which is characterized by the proliferation of nuclei in the loops of the glomeruli, and by consequent imperviousness of the same; in this way the flow of blood through the kidneys is greatly impaired or is even reduced to a minimum. The inexperienced observer may very easily overlook this important change completely, while the expert will have his attention at once directed to it through the contrast which the empty, and therefore large, glomeruli offer to the capillaries of the cortex and medulla when the latter are filled with blood.

As regards the origin of certain processes, we should always bear in mind the difficulties and complications that are present; we observe only that which has occurred, not that which is still in progress, and, without further evidence, a direct inference should not be drawn from the former concerning the latter.

Since the discovery of the power of locomotion of cells, the emigration of white blood-corpuscles, etc., observers have been accustomed to be as reserved as possible in this respect, so much the more as a direct practical interest does not usually attach to these questions. Some twenty years ago it was believed that we were much farther advanced in this matter than we are to-day; then even the beginner was always enjoined when studying tumors to first determine their "genesis." The purpose of this direction (which was really quite impracticable) was that the transitions of diseased into normal tissue should be studied, and even now examinations directed to this end are often advisable. Only do not imagine that just as soon as transitional forms have been determined the entire history of development has been discovered; in this way the most serious errors have been made.

# INDEX.

Abbé's apparatus, 2.
Aberration of light, 5.
Abscess, examination of pus from, 156.
   tubercle-bacilli in, 158.
Absolute alcohol, 24, 184.
Accessory apparatus, 8.
Acetate of potassium, 40.
Acetic acid, 18, 30.
Acid, acetic, 30.
   chromic, 34.
   nitric, as a decolorizer, 83.
   osmic, 89.
Acids, use of mineral, in decalcification, 28.
Actinomycetes, 158.
Albumin, coagulation of, by boiling, 93.
Albuminates, coagulation of, by alcohol, 25.
Alcohol, effect of, on specimens, 24.
Alkalies, 36.
Alum-carmine, 56.
Alum-cochineal solution, 57.
Amœboid movements, 116.
Ammonia-carmine, 51.
Ammonium sulphide, 91.
Amyloid substance, 48.
Anæmia, blood-corpuscles in, 128.
   mountaineers', 177.
*Anchylomum duodenale*, 5.
Angle of divergence, 177.
*Anguillula intestinalis*, 177.
Aniline-black, 62.
   -blue, 62.
   -dyes, 67.
Anthrax, bacillus of, 132.
Aqueous humor, 24.
Argand burner, 8.
Artificial digestion, 94.

Artificial gastric juice, 95.
   products, 19.
   serum, 24.
Asthma-crystals, 136.
Axis-cylinders of nerves, staining of, 52.

Bacilli, identification of, 68.
   of anthrax, 132.
   intestinal, 178.
   of leprosy, 75.
   of recurrent fever, 132.
   of typhus, 69, 74.
Bacilli of tubercle, 132, 147, 166, 179.
   importance of, 151, 153.
   identification when unstained, 69.
   staining of, 78.
Bacteria, in urine, 165.
Base, Schreiner's, 175.
Basic aniline dyes, 62.
Berlin-blue, 100, 101.
Bismarck-brown, 64, 77.
Blood, examination of, 126, 131.
   cellular elements in, 126.
   red corpuscles of, 128.
   white corpuscles of, 129.
   change in protoplasm of, 130.
   crystals of, 135, 136.
   parasites in, 131.
   anthrax-bacilli in, 132.
   spirilli in, 132, 133.
Blood-plates of Bizzozero, 127.
Blood-stains, examination of, 134.
Boiling of specimens, 93.
Borax-carmine, 56.
Buhl, theory of phthisis, 144.

Canada balsam, 40.
Capillaries, micrococci in, 70.
Carcinoma, cell-columns in, 171.
   diagnosis of, 170.

Carcinoma of uterus, 169.
Carmine, 51, 53.
  alum-c., 56.
  ammonia-c., 51.
  borax-c., 56.
  fluid-injection of, 102.
  lithium-c., 57.
Carmine-cement, 102.
Casts, 161.
  after fractures, 163.
  epithelial, 162.
  granular, 162.
  hyaline, 161.
  waxy, 162.
Catarrh, oral, cells in, 138.
Cedar, oil of, 6.
Celloidin, 99.
Cells, amœboid movements of, 116.
  decidual, 168.
  of endometrium, 168.
  endothelial, in blood, 131.
Cells, epithelial, in carcinoma, 170.
  in oral fluids, 138.
  in sputa, 141.
  in stools, 176.
  in urine, 163.
Cells, fatty degenerated in pus, 155.
  food-cells, 68.
Cells, mucus-, in stools, 176.
  in urine, 163.
Chloride of gold, 87.
Chloroform, 27.
Cholesterin, 51.
Chromic acid, 34.
Chyle, in exudations, 180.
Circulation, observation of, 109.
  in the frog's web, 110.
  in the tongue, 110.
  in the mesentery, 111.
  in the cornea, 112.
Clearing action of glycerin, 38.
  of Canada balsam, 40.
*Clostridium butyricum*, 178.
Cloves, oil of, 41.
Coagulation-necrosis, 65.
Cobalt-flasks for reagents, 18.
Cobbler's globe, 9.
Cochineal-alum solution, 57.
Concretions, intestinal, 176.
  in tonsils, 140.
Condenser, Abbé's, 3.
Connective tissue, action of acetic acid on, 32.
Cornea, circulation in, 112.
Corpora amylacea, 48.
Corpuscles, of blood, 127, 129.
  of pus, 155.

Corpuscles, salivary, 139.
  of syphilis, 127.
Correction-mountings, 6.
Cover-glasses, 9.
Crystals, asthma-, 136.
  hæmatoidin-, 136.
  hæmin-, 135.
  hæmoglobin-, 136.
  spermatic, 175.
Curschman's method of staining amyloid, 68.
Cylinder-diaphragm, 2.
Cysts, contents of, 179.

Decalcification, 28.
Degeneration, amyloid, 49, 50.
  fatty, 25.
  senile, 187.
Dehydration, with alcohol, 28.
Deposits of lime, 36.
  urinary, 159.
Detritus, significance of, in capillaries, 70.
Diabetes, glycogen in, 48.
Diagnosis, care necessary in making, 185.
Diaphragm, cylindrical, 2.
Digestion, artificial, 94.
Distilled water, 22.
*Distoma hæmatobium*, 132.
Double knife, 11.
Drawing, apparatus for, 7.
Dried preparations, Koch's method of staining, 121.
Drying of specimens, 94.
Dyeing, technique of, 43.
Dyes, aniline, 62, 66.
  carmine, 51, 53.
Dysmenorrhœa, pseudomembranous, 168.

Echinococci, in exudations, 182.
Egg-albumin in artificial serum, 24.
Election of dyes, 43.
Elements in tissue-fluids, 116.
Eosin, 131.
Eosinophil granules, 131.
Epithelium. See Cells, epithelial.
Epithelioma, true application of term, 172.
Erlitzki's hardening fluid, 35.
Erosion of cervix, distinguished from cancer, 169.
Errors in pathological histology, 186.
Ether, 27.
Entozoa in alimentary canal, 177.

# INDEX.   193

Extract of pancreas, 95.
Exudations, 177.
Eye-pieces, 7.

Fat, in exudations, 180.
  in pus-corpuscles, 155.
  removal of, 28.
  in urine after fractures, 163.
Fever, recurrent, spirilli in, 132.
  typhus, bacilli of, 74.
Fibers, elastic, in sputa, 145.
Fibrinous casts, 162.
*Filaria sanguinis*, 132.
Fluid, decalcifying, 29.
  examination of, 113.
  indifferent, 23, 24.
  Müller's, 34.
  suspended elements in, 114.
  tissue-, 117.
Formic acid, 33.
Food-cells, 66.
Forms of cells, difficulty in determining, 116.
Freezing specimens, 14.

Gangrene, sputum in, 145.
Gastric juice, artificial, 95.
Genital tract, secretions of, 166.
Glass apparatus, 9.
Globe, cobbler's, 9.
Glycerin, 38, 40.
Glycogen, 47, 181.
Gold, chloride of, 87.
Gonococcus, 173.
Gonorrhœa, secretions of, 173.
Gram's method of staining schizomycetes, 75.
Granules, eosinophil, 131.
Gum-arabic, 108.

Hæmatoblasts, 127.
Hæmatoidin, 136.
Hæmatoxylin, 58.
Hæmin, 135.
Hæmoglobin, 136.
Hardening in absolute alcohol, 24, 184.
  in chromic acid, 34.
  in gold chloride, 88.
  in osmic acid, 89.
  in picric acid, 33.
Hyaline casts, 161.
Hydrochloric acid, 28.
*Hydrops adiposus*, 180.
  *chylosus*, 180.

Identification of schizomycetes, 68.
  of tubercle-bacilli, 85.

Illuminating-lamp, 8.
Imbedding, 97.
  material used for, 98.
Immersion-lenses, 5, 6.
Indifferent fluid, 24.
Inflammation, microscopical observation of, 109.
Injecting, 100.
  apparatus for, 103.
  material used for, 100.
Instruments, metallic, 10.
Intestinal concretions, 176.
Intestine, parasites in, 177.
Iodine, 46.
  as a test for amyloid, 49.

Juice, artificial gastric, 95.

Leprosy, bacilli of, 75.
Leucocytes, increased number in blood, 129.
  granular protoplasm of, 130.
  in exudations, 180.
  movements of, in fluids, 116.
  in sputa, 141.
Leucocytosis, appearance of blood in, 130.
Light, for microscopic work, 8.
Lime, oxalate of, in urine, 160.
  phosphate of, 161.
Liquor potassæ, 36.
  sodæ, 36.
Lithium-carmine, 57.
Living tissues, observation of, 109.
Lugol's solution, 46.

Melanæmia, blood in, 131.
Mesentery, circulation in, 111.
Metallic instruments, 10.
Metals, noble, 86.
Menstrual fluid, 167.
Micro-chemical examinations, 21.
Micro-chemistry, 18.
Micrococci, in blood, 133.
  in blood-vessels, 70.
  in chains, 70.
  in gonorrhœal secretion, 173.
  identification of, 69.
  of pneumonia, 154.
  in pus, 158.
  in sputa, 139.
  in urine, 165.
Micro-organisms, examination of, 118.
  identification of, 68.
Microtome, 11, 16, 185.
Mouth, fluids of, 137.

Movement, amœboid, in cells, 116.
Mucous membrane of intestine, cells of, 176.
 of mouth, 138.
 of respiratory tract, 141.
 of uterus, 167.
Müller's fluid, 34.

Nitric acid, 28, 83.
Noble metals, 86.
Nuclei, staining of, 63.

Objectives, 4.
*Oidium albicans*, 139, 166.
Oil-immersion lenses, 6.
Oil of cloves, 40.
 of turpentine, 103.
Osmic acid, 89.
Oxalate of lime in urine, 160.
Oxaluria, 160.

Pancreas, extract of, 95.
Parasites, vegetable, in alimentary tract, 177.
Pathological discoveries, 186.
Phosphates in urine, 160.
Phthisis, diabetic, 153.
 tuberculous, 150.
 diagnosis of, 151.
 prognosis of, 153.
Picric acid, 33.
Picro-carmine, 54.
Picro-lithium-carmine, 57.
Pigment in lungs, 142.
Placenta, fragments of, 167.
Pneumococci, 154.
Potassium, acetate of, 21.
Pregnancy, uterine mucous membrane in, 168.
Preparations, dried, staining of, 121.
Preservation of specimens, 106.
Products, artificial, 19.
Prostatic secretion, 174.
Protoplasm, staining of, 130.
Pulmonary phthisis, 148.
Pus, actinomycetes in, 158.
 corpuscles of, 154.
 examination of, 155.
 foreign substances in, 156.
 fragments of bone in, 157.
 schizomycetes in, 158.

Razor, use of, 10.
Reagents, 18.
 for staining, 42.
Red blood-corpuscles, 91, 127.
Removal of fat, 27.

Salt-solution, 23.
*Sarcinæ ventriculi*, 177.
Schizomycetes in blood, 131.
 in oral fluids, 146.
 in pus, 158.
 staining of, 68.
 in urine, 165.
Schreiner's base, 175.
Sections, mounting of, 17.
 preservation of, 106.
 staining of, 43.
 thick and thin, 14.
Semen, 174.
Serum, artificial, 9.
Silver, 86.
Specimens, alcoholic, 23.
 boiling of, 93.
 preservation of, 106.
Spermatic crystals, 175.
Spermatozoa, 174.
Spirilli in recurrent fever, 132.
*Spirochæta Obermeyeri*, 132.
Splenic fever, bacilli of, 74.
Sputa, examination of, 136.
Staining, double, 61.
 dried preparations, 121.
 Gram's method, 75.
 of micrococci, 71.
 of nuclei, 63.
 principles of, 42.
 reagents used in, 46.
 of tubercle bacilli, 78.
Stand, microscopic, 2.
Stomach, contents of, 175.
Sulphide of ammonium, 91.
Sulphuric acid, 28.

Teichmann's test, 136.
Tissue-fluid, 117.
Tissues, living, observation of, 109.
Tongue, circulation in, 110.
*Trichomonas vaginalis*, 166.
Tubercle, bacillus of, 78, 147, 153, 158, 166, 179.
Tumor-elements, in exudations, 181.
 in urine, 164.
Tumors, examination of, 183.
Typhus, bacillus of, 74, 147.

Urates, 159.
Uric acid, 159.
Urine, casts in, 161.
 cells in, 163.
 examination of, 159.
 parasites in, 164.
 tumor-cells in, 164.

Uterus, cancer of, 170.
  erosion of, 169.
  fluids from, 167.

Vagina, secretion of, 166.
Vegetable parasites, 177.
Vital properties of cells, 114.

Water, abstraction of, 25.
  distilled, 22.
Web of frog, circulation in, 110.
Weigert's method of staining, 59.
White corpuscles of blood, 129.

Yeast-fungi, 177.

THE END.

JULY, 1885.

# MEDICAL
### AND
# HYGIENIC WORKS
PUBLISHED BY

*D. APPLETON & CO., 1, 3, & 5 Bond St., New York.*

BARKER (FORDYCE). On Sea-Sickness. Small 12mo. Cloth, 75 cents.
On Puerperal Disease. Third edition. 8vo. Cloth, $5.00; sheep, $6.00.
BARTHOLOW (ROBERTS). A Treatise on Materia Medica and Therapeutics. 8vo. Cloth, $5.00; sheep, $6.00.
A Treatise on the Practice of Medicine. 8vo. Cloth, $5.00; sheep, $6.00.
On the Antagonism between Medicines and between Remedies and Diseases. 8vo. Cloth, $1.25.
BASTIAN (H. CHARLTON). On Paralysis from Brain-Disease in its Common Forms. 12mo. Cloth, $1.75.
The Brain as an Organ of the Mind. 12mo. Cloth, $2.50.
BELLEVUE AND CHARITY HOSPITAL REPORTS. Edited by W. A. Hammond, M.D. 8vo. Cloth, $4.00.
BENNET (J. H.). Winter and Spring on the Shores of the Mediterranean. 12mo. Cloth, $3.50.
On the Treatment of Pulmonary Consumption, by Hygiene, Climate, and Medicine. Thin 8vo. Cloth, $1.50.
BILLINGS (F. S.). The Relation of Animal Diseases to the Public Health, and their Prevention. 8vo. Cloth, $4.00.
BILLROTH (THEODOR). General Surgical Pathology and Therapeutics. 8vo. Cloth, $5.00; sheep, $6.00.
BRAMWELL (BYROM). Diseases of the Heart and Thoracic Aorta. 8vo. Cloth, $8.00; sheep, $9.00.
BUCK (GURDON). Contributions to Reparative Surgery. 8vo. Cloth, $3.00.
CARPENTER (W. B.). Principles of Mental Physiology, with their Application to the Training and Discipline of the Mind, and the Study of its Morbid Conditions. 12mo. Cloth, $3.00.
CARTER (ALFRED H.). Elements of Practical Medicine. 12mo. Cloth, $3.00.
CHAUVEAU (A.). The Comparative Anatomy of the Domesticated Animals. 8vo. Cloth, $6.00.
COMBE (ANDREW). The Management of Infancy, Physiological and Moral. 12mo. Cloth, $1.50.
COOLEY. Cyclopædia of Practical Receipts, and Collateral Information in the Arts, Manufactures, Professions, and Trades, including Medicine, Pharmacy, and Domestic Economy. 2 vols., 8vo. Cloth, $9.00.
CORNING (J. L.). Brain Exhaustion, with some Preliminary Considerations on Cerebral Dynamics. Crown 8vo. Cloth, $2.00.
DAVIS (HENRY G.). Conservative Surgery. 8vo. Cloth, $3.00.
ELLIOT (GEORGE T.). Obstetric Clinic. 8vo. Cloth, $4.50.
EVETZKY (ETIENNE). The Physiological and Therapeutical Action of Ergot. 8vo. Limp cloth, $1.00.
FLINT (AUSTIN, JR.). The Physiological Effects of Severe and Protracted Muscular Exercise; with Special Reference to its Influence upon the Excretion of Nitrogen. 12mo. Cloth, $1.00.
Text-Book of Human Physiology. Imperial 8vo. Cloth, $6.00; sheep, $7.00.

MEDICAL AND HYGIENIC WORKS.—(*Continued.*)

FLINT (AUSTIN, JR.). The Source of Muscular Power. 12mo. Cloth, $1.00.
Manual Chemical Examinations of the Urine in Disease. 12mo. Cloth, $1.00.
FOURNIER (ALFRED). Syphilis and Marriage. 8vo. Cloth, $2.00; sheep, $3.00.
FREY (HEINRICH). The Histology and Histochemistry of Man. 8vo. Cloth, $5.00; sheep, $6.00.
FRIEDLANDER (CARL). The Use of the Microscope in Clinical and Pathological Examinations. 8vo. (*Nearly ready.*)
GAMGEE (JOHN). Yellow Fever a Nautical Disease. 8vo. Cloth, $1.50.
GROSS (SAMUEL W.). A Practical Treatise on Tumors of the Mammary Gland. 8vo. Cloth, $2.50.
GUTMANN (EDWARD). The Watering-Places and Mineral Springs of Germany, Austria, and Switzerland. 12mo. Cloth, $2.50.
GYNÆCOLOGICAL TRANSACTIONS, VOL. VIII. Being the Proceedings of the Eighth Annual Meeting of the American Gynæcological Society, held in Philadelphia, September 18, 19, and 20, 1883. 8vo. Cloth, $5.00.
GYNÆCOLOGICAL TRANSACTIONS, VOL. IX. Being the Proceedings of the Ninth Annual Meeting of the American Gynæcological Society, held in Chicago, September 30, and October 1 and 2, 1884. 8vo. Cloth, $5.00.
HAMILTON (ALLAN McL.). Clinical Electro-Therapeutics, Medical and Surgical. 8vo. Cloth, $2.00.
HAMMOND (W. A.). A Treatise on Diseases of the Nervous System. 8vo. Cloth, $5.00; sheep, $6.00.
A Treatise on Insanity, in its Medical Relations. 8vo. Cloth, $5.00; sheep, $6.00.
Clinical Lectures on Diseases of the Nervous System. 8vo. Cloth, $3.50.
HART (D. BERRY). Atlas of Female Pelvic Anatomy. Large 4to. (*Sold only by subscription.*) Cloth, $15.00.
HARVEY (A.). First Lines of Therapeutics. 12mo. Cloth, $1.50.
HEALTH PRIMERS. In square 16mo volumes. Cloth, 40 cents each.
  I. Exercise and Training.
  II. Alcohol: Its Use and Abuse.
  III. Premature Death: Its Promotion or Prevention.
  IV. The House and its Surroundings.
  V. Personal Appearance in Health and Disease.
  VI. Baths and Bathing.
  VII. The Skin and its Troubles.
  VIII. The Heart and its Functions.
  IX. The Nervous System.
HOFFMANN-ULTZMANN. Introduction to an Investigation of Urine, with Special Reference to Diseases of the Urinary Apparatus. 8vo. Cloth, $2.00.
HOWE (JOSEPH W.). Emergencies, and how to treat them. 8vo. Cloth, $2.50.
The Breath, and the Diseases which give it a Fetid Odor. 12mo. Cloth, $1.00.
HUXLEY (T. H.). The Anatomy of Vertebrated Animals. 12mo. Cloth, $2.50.
The Anatomy of Invertebrate Animals. 12mo. Cloth, $2.50.
JACCOUD (S.). The Curability and Treatment of Pulmonary Phthisis. 8vo. Cloth, $4.00.
JOHNSON (JAMES F. W.). The Chemistry of Common Life. 12mo. Cloth, $2.00.
JONES (H. MACNAUGHTON). Practical Manual of Diseases of Women and Uterine Therapeutics. 12mo. Cloth, $3.00.
KEYES (E. L.). The Tonic Treatment of Syphilis, including Local Treatment of Lesions. 8vo. Cloth, $1.00.
KINGSLEY (N. W.). A Treatise on Oral Deformities as a Branch of Mechanical Surgery. 8vo. Cloth, $5.00; sheep, $6.00.
LEGG (J. WICKHAM). On the Bile, Jaundice, and Bilious Diseases. 8vo. Cloth, $6.00; sheep, $7.00.

MEDICAL AND HYGIENIC WORKS.—(Continued.)

LETTERMANN (JONATHAN). Medical Recollections of the Army of the Potomac. 8vo. Cloth, $1.00.
LITTLE (W. J.). Medical and Surgical Aspects of In-Knee (Genu-Valgum): Its Relation to Rickets, its Prevention, and its Treatment, with and without Surgical Operation. 8vo. Cloth, $2.00.
LORING (EDWARD G.). A Text-Book of Ophthalmoscopy. Part I. The Normal Eye, Determination of Refraction, and Diseases of the Media. 8vo. (In press.)
LUSK (WILLIAM T.). The Science and Art of Midwifery. 8vo. Cloth, $5.00; sheep, $6.00.
LUYS (J.). The Brain and its Functions. 12mo. Cloth, $1.50.
McSHERRY (RICHARD). Health, and how to promote it. 12mo. Cloth, $1.25.
MARKOE (T. M.). A Treatise on Diseases of the Bones. 8vo. Cloth, $4.50.
MAUDSLEY (HENRY). Body and Mind: An Inquiry into their Connection and Mutual Influence, specially in reference to Mental Disorders. 12mo. Cloth, $1.50.
— Physiology of the Mind. 12mo. Cloth, $2.00.
— Pathology of the Mind. 12mo. Cloth, $2.00.
— Responsibility in Mental Disease. 12mo. Cloth, $1.50.
NEUMANN (ISIDOR). Hand-Book of Skin Diseases. 8vo. Cloth, $4.00; sheep, $5.00.
THE NEW YORK MEDICAL JOURNAL (weekly). Edited by Frank P. Foster, M. D. Terms per annum, $5.00.
GENERAL INDEX, from April, 1865, to June, 1876 (23 volumes). 8vo. Cloth, 75 cents.
NIEMEYER (FELIX VON). A Text-Book of Practical Medicine, with particular reference to Physiology and Pathological Anatomy. 2 volumes, 8vo. Cloth, $9.00; sheep, $11.00.
NIGHTINGALE'S (FLORENCE) Notes on Nursing. 12mo. Cloth, 75 cents.
OSWALD (F. L.). Physical Education; or, The Health Laws of Nature. 12mo. Cloth, $1.00.
PEASLEE (E. R.). A Treatise on Ovarian Tumors: Their Pathology, Diagnosis, and Treatment, with reference especially to Ovariotomy. 8vo. Cloth, $5.00; sheep, $6.00.
PEREIRA'S (DR.) Elements of Materia Medica and Therapeutics. Royal 8vo. Cloth, $7.00; sheep, $8.00.
PEYER (ALEX.). Clinical Microscopy. Translated by A. C. Girard, M. D. (Nearly ready.)
POMEROY (OREN D.). The Diagnosis and Treatment of Diseases of the Ear. 8vo. (Nearly ready.)
POORE (C. T.). Osteotomy and Osteoclasis, for the Correction of Deformities of the Lower Limbs. 8vo. Cloth, $2.50.
QUAIN (RICHARD). A Dictionary of Medicine, including General Pathology, General Therapeutics, Hygiene, and the Diseases peculiar to Women and Children. By Various Writers. Edited by Richard Quain, M. D. 8vo. (Sold only by subscription.) Half morocco, $8.00.
RANNEY (AMBROSE L.). Applied Anatomy of the Nervous System. 8vo. Cloth, $4.00; sheep, $5.00.
RIBOT (TH.). Diseases of Memory: An Essay in the Positive Psychology. 12mo. Cloth, $1.50.
RICHARDSON (B. W.). Diseases of Modern Life. 12mo. Cloth, $2.00.
— A Ministry of Health and other Addresses. 12mo. Cloth, $1.50.
ROBINSON (A. R.). A Manual of Dermatology. 8vo. Cloth, $5.00.
ROSENTHAL (I.). General Physiology of Muscles and Nerves. 12mo. Cloth, $1.50.

MEDICAL AND HYGIENIC WORKS.—(Continued.)

ROSCOE AND SCHORLEMMER. Treatise on Chemistry.
Vol. 1. Non-Metallic Elements. 8vo. Cloth, $5.00.
Vol. 2. Part I. Metals. 8vo. Cloth, $3.00.
Vol. 2. Part II. Metals. 8vo. Cloth, $3.00.
Vol. 3. Part I. The Chemistry of the Hydrocarbons and their Derivatives. 8vo. Cloth, $5.00.
Vol. 3. Part II. The Chemistry of the Hydrocarbons and their Derivatives. 8vo. Cloth, $5.00.

SAYRE (LEWIS A.). Practical Manual of the Treatment of Club-Foot. 12mo. Cloth, $1.25.
Lectures on Orthopedic Surgery and Diseases of the Joints. 8vo. Cloth, $5.00; sheep, $6.00.

SCHROEDER (KARL). A Manual of Midwifery, including the Pathology of Pregnancy and the Puerperal State. 8vo. Cloth, $3.50; sheep, $4.50.

SIMPSON (JAMES Y.). Selected Works: Anæsthesia, Diseases of Women. 3 volumes, 8vo. Per volume, cloth, $3.00; sheep, $4.00.

SMITH (EDWARD). Foods. 12mo. Cloth, $1.75.
Health. 12mo. Cloth, $1.00.

STEINER (JOHANNES). Compendium of Children's Diseases. 8vo. Cloth, $3.50; sheep, $4.50.

STONE (R. FRENCH). Elements of Modern Medicine, including Principles of Pathology and of Therapeutics, with many Useful Memoranda and Valuable Tables of Reference. Accompanied by Pocket Fever Charts. Wallet-book form, with pockets on each cover for Memoranda, Temperature Charts, etc. (In press.)

STRECKER (ADOLPH). Short Text-Book of Organic Chemistry. 8vo. Cloth, $5.00.

SWANZY (HENRY R.). A Hand-Book of the Diseases of the Eye, and their Treatment. Crown 8vo. Cloth, $3.00.

TRACY (ROGER S.). The Essentials of Anatomy, Physiology, and Hygiene. 12mo. Cloth, $1.25.
Hand-Book of Sanitary Information for Householders. 16mo. Cloth, 50 cents.

TRANSACTIONS OF THE NEW YORK STATE MEDICAL ASSOCIATION, VOLUME I. Being the Proceedings of the First Annual Meeting of the New York State Medical Association, held in New York, November 18, 19, and 20, 1884. Small 8vo. Cloth, $5.00.

TYNDALL (JOHN). Essays on the Floating Matter of the Air, in Relation to Putrefaction and Infection. 12mo. Cloth, $1.50.

ULTZMANN (ROBERT). Pyuria, or Pus in the Urine, and its Treatment. 12mo. Cloth, $1.00.

VAN BUREN (W. H.). Lectures upon Diseases of the Rectum, and the Surgery of the Lower Bowel. 8vo. Cloth, $3.00; sheep, $4.00.
Lectures on the Principles and Practice of Surgery. 8vo. Cloth, $4.00; sheep, $5.00.

VAN BUREN AND KEYES. A Practical Treatise on the Surgical Diseases of the Genito-Urinary Organs, including Syphilis. 8vo. Cloth, $5.00; sheep, $6.00.

VOGEL (A.). A Practical Treatise on the Diseases of Children. 8vo. Cloth, $4.50; sheep, $5.50.

WAGNER (RUDOLF). Hand-Book of Chemical Technology. 8vo. Cloth, $5.00.

WALTON (GEORGE E.). Mineral Springs of the United States and Canadas. 12mo. Cloth, $2.00.

WEBBER (S. G.). A Treatise on Nervous Diseases. 8vo. (Nearly ready.)

WELLS (T. SPENCER). Diseases of the Ovaries. 8vo. Cloth, $4.50.

WYLIE (WILLIAM G.). Hospitals: Their History, Organization, and Construction. 8vo. Cloth, $2.50.

# A DICTIONARY OF MEDICINE, including General Pathology, General Therapeutics, Hygiene, and the Diseases peculiar to Women and Children. By Various Writers.

### Edited by RICHARD QUAIN, M. D., F. R. S.,

Fellow of the Royal College of Physicians; Member of the Senate of the University of London; Member of the General Council of Medical Education and Registration; Consulting Physician to the Hospital for Consumption and Diseases of the Chest at Brompton, etc.

In one large 8vo volume of 1,834 pages, and 138 Illustrations. Half morocco, $8.00. Sold only by subscription.

This work is primarily a Dictionary of Medicine, in which the several diseases are fully discussed in alphabetical order. The description of each includes an account of its etiology and anatomical characters; its symptoms, course, duration, and termination; its diagnosis, prognosis, and, lastly, its treatment. General Pathology comprehends articles on the origin, characters, and nature of disease.

General Therapeutics includes articles on the several classes of remedies, their modes of action, and on the methods of their use. The articles devoted to the subject of Hygiene treat of the causes and prevention of disease, of the agencies and laws affecting public health, of the means of preserving the health of the individual, of the construction and management of hospitals, and of the nursing of the sick.

Lastly, the diseases peculiar to women and children are discussed under their respective headings, both in aggregate and in detail.

Among the leading contributors, whose names at once strike the reader as affording a guarantee of the value of their contributions are the following:

ALLBUTT, T. CLIFFORD, M. A., M. D.
BARNES, ROBERT, M. D.
BASTIAN, H. CHARLTON, M. A., M. D.
BINZ, CARL, M. D.
BRISTOWE, J. SYER, M. D.
BROWN-SÉQUARD, C. E., M.D., LL.D.
BRUNTON, T. LAUDER, M. D., D. Sc.
FAYRER, Sir JOSEPH, K. C. S. I., M. D., LL. D.
FOX, TILBURY, M. D.
GALTON, Captain DOUGLAS, R. E. (retired).
GOWERS, W. R., M. D.
GREENFIELD, W. S., M. D.
JENNER, Sir WILLIAM, Bart., K.C.B., M. D.
LEGG, J. WICKHAM, M. D.
NIGHTINGALE, FLORENCE.
PAGET, Sir JAMES, Bart.
PARKES, EDMUND A., M. D.
PAVY, F. W., M. D.
PLAYFAIR, W. S., M. D.
SIMON, JOHN, C. B., D. C. L.
THOMPSON, Sir HENRY.
WATERS, A. T. H., M. D.
WELLS, T. SPENCER.

"Not only is the work a Dictionary of Medicine in its fullest sense; but it is so encyclopedic in its scope that it may be considered a condensed review of the entire field of practical medicine. Each subject is marked up to date and contains in a nutshell the accumulated experience of the leading medical men of the day. As a volume for ready reference and careful study, it will be found of immense value to the general practitioner and student."— *Medical Record.*

New York: D. APPLETON & CO., 1, 3, & 5 Bond Street.

THE
# NEW YORK MEDICAL JOURNAL,
### A WEEKLY REVIEW OF MEDICINE.

Edited by FRANK P. FOSTER, M. D.

## THE LEADING JOURNAL OF AMERICA.

---

Containing twenty-eight double-columned pages of reading-matter, consisting of **Lectures, Original Communications, Clinical Reports, Correspondence, Book Notices, Leading Articles, Minor Paragraphs, News Items, Letters to the Editor, Proceedings of Societies, Reports on the Progress of Medicine,** and **Miscellany.**

By reason of the condensed form in which the matter is arranged, the JOURNAL contains more reading-matter than any other of its class in the United States. Its pages contain an average of 1,300 words; each volume has at least 748 pages, giving an aggregate of 972,400 words, or more than double the amount of reading-matter contained in a $5.00 octavo volume of 800 pages, averaging 500 words to the page. It is also more freely illustrated, and its illustrations are generally better executed, than is the case with other weekly journals.

The articles contributed to the JOURNAL are of a high order of excellence, for authors know that through its columns they address the better part of the profession; a consideration which has not escaped the notice of advertisers, as shown by its increasing advertising patronage.

---

The volumes begin with January and July of each year. Subscriptions can be arranged to begin with the volume.

### TERMS, PAYABLE IN ADVANCE.

One Year - - - - - - $5 00
Six Months - - - - - - - 2 50

The Popular Science Monthly and The New York Medical Journal to the same address, $9.00 per Annum (full price, $10.00).

---

New York: D. APPLETON & CO., 1, 3, & 5 Bond Street.

# THE POPULAR SCIENCE MONTHLY.

### CONDUCTED BY E. L. AND W. J. YOUMANS.

The Popular Science Monthly will continue, as heretofore, to supply its readers with the results of the latest investigation and the most valuable thought in the various departments of scientific inquiry.

Leaving the dry and technical details of science, which are of chief concern to specialists, to the journals devoted to them, the Monthly deals with those more general and practical subjects which are of the greatest interest and importance to the public at large. In this work it has achieved a foremost position, and is now the acknowledged organ of progressive scientific ideas in this country.

The wide range of its discussions includes, among other topics:

The bearing of science upon education;

Questions relating to the prevention of disease and the improvement of sanitary conditions;

Subjects of domestic and social economy, including the introduction of better ways of living, and improved applications in the arts of every kind;

The phenomena and laws of the larger social organizations, with the new standard of ethics, based on scientific principles;

The subjects of personal and household hygiene, medicine, and architecture, as exemplified in the adaptation of public buildings and private houses to the wants of those who use them;

Agriculture and the improvement of food-products;

The study of man, with what appears from time to time in the departments of anthropology and archæology that may throw light upon the development of the race from its primitive conditions.

Whatever of real advance is made in chemistry, geography, astronomy, physiology, psychology, botany, zoölogy, paleontology, geology, or such other department as may have been the field of research, is recorded monthly.

Special attention is also called to the biographies, with portraits, of representative scientific men, in which are recorded their most marked achievements in science, and the general bearing of their work indicated and its value estimated.

Terms: $5.00 per annum, in advance.
The New York Medical Journal and The Popular Science Monthly to the same address, $9.00 per annum (full price, $10.00).

New York: D. APPLETON & CO., 1, 3, & 5 Bond Street.

# RECENT PUBLICATIONS.

**A Hand-Book of the Diseases of the Eye, and their Treatment.** By HENRY R. SWANZY, A. M., M. B., F. R. C. S. I., Surgeon to the National Eye and Ear Infirmary; Ophthalmic Surgeon to the Adelaide Hospital, Dublin. Crown 8vo, 437 pages. With 122 Illustrations, and Holmgren's Tests for Color-Blindness. $3.00.

"The above is a handy manual, intended specially for students about to undertake the study of the eye, and it has been arranged specially with the view of being useful to them in carrying out such studies systematically. Subjects which to beginners are found particularly embarrassing or difficult are thus given prominence, and receive careful attention. It contains the essentials without redundant matter, and seems admirably suited to the end in view."—*Canada Medical and Surgical Journal.*

"This hand-book is intended for students in ophthalmology, and is among the best in this specialty that we have perused. The author is a clear writer and practical ophthalmologist, and gives here very practical instruction in this important department of medical science."—*Buffalo Medical and Surgical Journal.*

**Elements of Practical Medicine.** By ALFRED H. CARTER, M. D., Member of the Royal College of Physicians, London; Physician to the Queen's Hospital, Birmingham, etc. Third edition, revised and enlarged. 1 vol., 12mo, cloth, $3.00.

"Although this work does not profess to be a complete treatise on the practice of medicine, it is too full to be called a compend; it is rather an introduction to the more exhaustive study embodied in the larger text-books. Notwithstanding the condensed make-up of the book, it is quite comprehensive, including even cutaneous and venereal diseases. It contains much valuable information, and we may add that it is very readable."—*New York Medical Journal.*

**Osteotomy and Osteoclasis, for the Correction of Deformities of the Lower Limbs.** By CHARLES T. POORE, M. D., Surgeon to St. Mary's Free Hospital for Children, New York. 1 vol., 8vo, 187 pages, 50 Illustrations, cloth, $2.50.

"There has been a want of a concise treatise on osteotomy, and the author aims in this work to supply the void. He has succeeded fully in his object, and presents to the profession, for their guidance in the treatment of a very common class of deformities, a very valuable work. . . . The subjects and diseases treated are such as the surgeon is compelled to consider and examine often in his daily practice, and the author has drawn from his own ample experience and that of other eminent surgeons in the preparation of a most admirable and useful work."—*Buffalo Medical and Surgical Journal.*

"This is an interesting and practical monograph, in which the author has admirably succeeded in laying before the medical public the operative procedures usually resorted to for the correction of osseous deformities of the lower extremities. The author's style is clear and forcible, the illustrations are well executed and to the point, and the typographical part is creditable to the publishers. The work will be a useful addition to the surgeon's library."—*St. Louis Medical and Surgical Journal.*

New York: D. APPLETON & CO., 1, 3, & 5 Bond Street.

# RECENT PUBLICATIONS.

**Practical Manual of Diseases of Women and Uterine Therapeutics.** For Students and Practitioners. By H. MACNAUGHTON JONES, M. D., F. R. C. S. I. and E., Examiner in Obstetrics, Royal University of Ireland; Fellow of the Academy of Medicine in Ireland; and of the Obstetrical Society of London, etc. 1 vol., 12mo, 410 pages, with 188 Illustrations, cloth, $3.00.

"Within a very moderate compass, this work covers the field of gynæcology quite fully, but it deals most prominently with the minor details of that branch—what may be termed every-day gynæcology. Methods of diagnosis are presented with much completeness, and so also are measures of treatment, including, as is apt not to be the case with books of this class, medicinal treatment. For these reasons, as well as because the author's views will be found to be eminently judicious, we think that the book deserves to be widely studied by general practitioners and students."—*The New York Medical Journal.*

"Those who desire to obtain, at a minimum cost of time and money, a better acquaintance than the present educational facilities of this country present to the aspirants for gynæcological celebrity, will find in this work of Dr. Jones, conveyed in clear and plain terms, if not all that the modern infinitude of female diseases may seem to demand, yet perhaps sufficient to serve their more pressing needs, not only in the line of positive instruction, but also in that which is not less useful to the ambitious neophyte—salutary admonition."—*Canada Lancet.*

**Atlas of Female Pelvic Anatomy.** By D. BERRY HART, M. D., F. R. C. P. E., Lecturer on Midwifery and Diseases of Women, School of Medicine, Edinburgh, etc. With Preface by ALEXANDER J. C. SKENE, M. D., Professor of the Medical and Surgical Diseases of Women, Long Island College Hospital, Brooklyn, etc. Large 4to, 37 Plates with 150 Figures, and 89 pages descriptive text. Cloth, $15.00.

"Within recent years much has been done to weed the topographical anatomy of the pelvis of numerous errors which have encumbered it. Prominent among those who have furthered this work is the author of the 'Atlas' now before us, and into this, his latest labor, he has entered with all his accustomed vigor. . . . The 'Atlas' deserves, and will surely have, a wide circulation; and we are confident that no one will rise from its careful perusal without having obtained clearer, more accurate, and more intelligent views in regard to the much-vexed questions of female anatomy, or without having formed a very high opinion of the author's industry, earnestness, and ability."—*Edinburgh Medical Journal.*

"As an exposition of anatomical details, in their relation to obstetrics and gynæcology, it has no rival in the English language, and we may predict that it will, for years to come, occupy the position of a standard work of reference on these subjects."—*Glasgow Medical Journal.*

**The Curability and Treatment of Pulmonary Phthisis.** By S. JACCOUD, Professor of Medical Pathology to the Faculty of Paris; Member of the Academy of Medicine; Physician to the Lariboisière Hospital, Paris, etc. Translated and edited by MONTAGU LUBBOCK, M. D. (London and Paris), M. R. C. P. (England), etc. 8vo, 407 pages, cloth, $4.00.

New York: D. APPLETON & CO., 1, 3, & 5 Bond Street.

# RECENT PUBLICATIONS.

**A Manual of Dermatology.** By A. R. ROBINSON, M. B., L. R. C. P. and S. (Edin.), Professor of Dermatology at the New York Polyclinic; Professor of Histology and Pathological Anatomy at the Woman's Medical College of the New York Infirmary. Revised and corrected. 8vo, 647 pages, cloth, $5.00.

"It includes so much good, original work, and so well illustrates the best practical teachings of the subject by our most advanced men, that I regard it as commanding at once a place in the very front rank of all authorities. . . ."—JAMES NEVINS HYDE, M. D.

"Dr. Robinson's experience has amply qualified him for the task which he assumed, and he has given us a book which commends itself to the consideration of the general practitioner."—*Medical Age.*

"In general appearance it is similar to Duhring's excellent book, more valuable, however, in that it contains much later views, and also on account of the excellence of the anatomical description accompanying the microscopical appearances of the diseases spoken of."—*St. Louis Medical and Surgical Journal.*

"Altogether it is an excellent work, helpful to every one who consults its pages for aid in the study of skin-diseases. No physician who studies it will regret the placing of it in his library.—*Detroit Lancet.*

## NEARLY READY.—June, 1885.

**The Use of the Microscope in Clinical and Pathological Examinations.** By Dr. CARL FRIEDLAENDER, Docent in Pathological Anatomy at Berlin. Second edition, enlarged and improved, with a chromo-lithograph Plate. Translated, with the permission of the author, by HENRY C. COE, M. D., M. R. C. S., L. R. C. P. (London), Pathologist to the Woman's Hospital of the State of New York. 8vo vol. of 200 pages.

**The Diagnosis and Treatment of Diseases of the Ear.** By OREN D. POMEROY, M. D., Surgeon to the Manhattan Eye and Ear Hospital, etc. With one hundred Illustrations. New edition, revised and enlarged. 8vo, cloth.

**Clinical Microscopy.** By Dr. ALEX. PEYER. New edition, revised and enlarged. With ninety Plates, comprising one hundred and five Illustrations. Translated by A. C. GIRARD, M. D., Assistant Surgeon, U. S. Army.

**A Text-Book of Ophthalmoscopy.** By EDWARD G. LORING, M. D. Part I.—The Normal Eye, Determination of Refraction, and Diseases of the Media. 8vo. Profusely illustrated.

New York: D. APPLETON & CO., 1, 3, & 5 Bond Street.

-product-compliance